How to Write a Usable User Manual

The Professional Writing Series

This volume is one of a series published by ISI Press®. The series is designed to improve the communication skills of professional men and women, as well as students embarking upon professional careers.

Books published in this series:

How to Write a Usable User Manual

Edmond H. Weiss

iSi PRESS®
Philadelphia

Published by

iSi PRESS® A Subsidiary of the
Institute for Scientifc Information®
3501 Market Street, Philadelphia, PA 19104 U.S.A.

© 1985 ISI Press

Library of Congress Cataloging in Publication Data

Weiss, Edmond H.
  How to write a usable user manual.

  Bibliography: p.
  Includes index.
  1. Electronic data processing – Authorship.
2. Computers – Handbooks, manuals, etc. 3. Technical
writing. I. Title. II. Series.
QA76.165.W45 1985   808'.066004   85-14429
ISBN 0-89495-051-7
ISBN 0-89495-052-5 (pbk.)

Printed in the United States of America
92 91 90 89 88 87   8 7 6 5 4 3

For Beverly

Contents

Part III   The Future of User Documentation

Appendixes

Preface: Who Needs This Book

This book is for anyone responsible for at least one user manual. It is for all the people who must plan, write, edit, or publish user documentation—and also for the people who manage them. More important, though, it is for those who are generally unhappy with most of the user documentation they have seen—and even with some they have written. For these readers, this book offers a set of ideas, principles, and techniques with which user manuals can be made more suitable, accessible, and readable.

These are interesting times for user documentation. Writing user documentation—especially software manuals for users and operators—is the subject of several new books and graduate courses. It is emerging as a distinct profession, incorporating elements of computer science, training, graphic arts, and traditional principles of rhetoric and composition.

Most of the causes of this trend are apparent. In just the past ten years, the numbers of computers, software products, and associated peripheral devices have exploded. And, almost without exception, each of these tens of thousands of systems, instruments, and gadgets has needed some literature for its users and owners.

Even more important than the increase in the sheer quantity of material has been the profound change in the kinds of audiences for whom the material is written. Once, only a decade ago, the typical reader of an operations manual was a scientist, engineer, or skilled technician. Today, however, the typical reader is a business professional with little or no training in computer technology, or a clerk or technician who cannot be expected to use the difficult and complicated manuals that were commonplace in the mid-1970s.

Although there have always been some good and even excellent user manuals, they have been quite rare and exceptional. Previously, though, when computer operators were mainly engineers and highly trained technicians, there was less need for accessible and easy to read user documentation. People in the technical professions expect manuals to be difficult, and they usually have learned to fight their way through tangled publications. Today, though, the typical users of a computer system either will not battle with an unfriendly manual or, in many cases, cannot.

Interestingly, the fundamental requirements for good user documentation have not changed much over the years. Manuals must still be **available on time**. They must be as **complete and accurate** as review and testing can make them. They must be written in **correct language** and edited for **clarity and readability**.

What has changed, though, are the tolerances and standards. Today's users – even the expert technicians and engineers – have less and less patience with unreadable and inaccessible publications. The old joke – when all else fails, read the instructions – is no longer funny. Finally, manufacturers of systems and software have realized that effective user documentation is often the key to success in a competitive market. Operations supervisors are at last understanding how much more productive a data processing center can be with the right manuals and guides. And there are even some analysts who have learned that writing user documentation early in the life of an application tends to improve its quality.

But just wanting to write better user manuals is not enough. Even today, most user documentation is written by people without any special training. I cannot be sure, but I suspect that the typical user manual is still the work of an analyst or programmer who has been drafted for the job. Many other publications are written by talented people recently recruited from teaching and the humanities, who, although they are an extremely salutary addition to the computer industry, have little preparation in the craft of user documentation. Indeed, even many long-time publications professionals complain that the old, one-size-fits-all documentation standards from the 1960s are no longer working.

So, the goal of this book is to help all these groups – and any others who want a set of principles for planning, designing, and writing user documentation. For people new to the field, it is meant to be a useful kit of ideas for attacking a job that often intimidates the newcomer: explaining a complicated product or system to a naive, or sometimes nontechnical, audience. For experienced technical writers and editors, for publications professionals, it is meant as a tool for analyzing how user documentation can be improved and for justifying the cost of those improvements to skeptical associates and superiors. What I hope will be most interesting and practical about *How to Write a Usable User Manual* is its central theme: **That developing a set of user manuals is very much like developing a complicated computer program.**

NOTE: This book is presented in a relatively unusual format and style. More precisely, it is in the format and style I recommend for many user manuals. Of course, this book is *not* a user manual, and, as a result, this style may sometimes seem odd or inappropriate. The publishers and I have taken this risk deliberately, believing that in user documentation, as in most things, the best teacher is still example.

Acknowledgments

Countless people have helped to form the ideas in this book. I have profited from the published writings of many software and documentation specialists. And I have learned much from my clients, who taught me what the problems are. In particular I must thank Stephen Bean, Kenneth Helms, Madeline Flynn, and Michael Bartlett of the NCR Corporation, who not only asked questions but also provided several of the answers.

Small portions of this book appeared in altered form in *Data Training* newspaper. Excerpts from Chapter 2 appeared in the February 1984 issue and excerpts from Chapter 8 appeared in June 1983 issue. These are used with the permission of DATA TRAINING, Warren/Weingarten, Inc., 38 Chauncy St, Boston MA 02111, (617) 542-0146.

For permission to reproduce copyrighted excerpts from company user manuals, I wish to acknowledge the following:

Jennie Balcom, Pioneer Data Systems, *VAX MAIL User's Guide*
McDonnell Douglas Computer Systems Company, *Introduction to REALITY*
Wharton Econometric Forecasting Associates, Inc., Philadelphia, PA, *Tables Documentation: A Teaching Introduction*

I am also one of a small group of especially fortunate authors who have been guided by Maryanne Soper of ISI Press.

PART

I

Toward a Science
of User Documentation

Chapter

1

Functions: What User
Manuals Do

1.1   What Is a User? What Is a Manual?

Users are people who must be *satisfied*. Organizations or individuals buy and develop computer technology with some goals, in pursuit of particular advantages. Generally, the users are the ones who must be convinced that the goals have been met and the advantages realized. User manuals are information products that help these users get the fullest value and benefit possible.

Any definition of *user* is risky. With appropriate caution, then, I'll define a *user* as **a person more concerned with the *outcome* of data processing than with the *output*.**

When I say that users are people who must be *satisfied*, I mean that they must believe the objectives have been met, and with acceptable effort and cost. The end matters more than the means; the outcome more than the output. How the system works is less important than how to work the system to the advantage and benefit of the users.

Surely, users may become interested in the inner workings of computer technology. But this does not alter the basic idea: Users are people who want something bigger than, and outside of, the particular device they are using. If they could find a cost-effective way to get what they want without computer technology, they would.

Why is it necessary to stress this point? Because so many of the people who develop computer technology—and the associated documentation—tend to view the technology as an end in itself. And because the ensuing user manuals are so often unusable technical treatises about the product, rather than tools to help the users get what they really want and need.

For bigger systems and products, the users are often entire organizations, with specialized interests and skills: corporate executives, interested only in the reports; functional managers, seeking administrative support; senior operators, charged with keeping the system going; junior operators and clerks, feeding the system data and monitoring its performance reports; maintenance technicians and programmers; auditors and quality assurance specialists.

Again, what all these diverse groups have in common is that they must be *satisfied* by the people who develop systems and programs. For those who sell computer products in the marketplace, the users are *customers*. For those who develop systems and applications within their own organizations, the users are *the managers of the functional departments*.

A user manual is—or should be—a tool that helps its readers get full benefit and advantage from the system. Traditionally, user manuals compensate for the difficulty and unfriendliness in computer technology; they answer such questions as: What do I do next? What does this mean? What is it doing now? Why didn't that work?

EXHIBIT 1.1   Main Uses of User Documentation

Function	For Example . . .
Help the User Get Started	• introduce functions and benefits • demonstrate installation and setup • teach elementary operations/ procedures • warn against errors and bugs
Help Productivity/Satisfaction	• demonstrate advanced features/ benefits • teach more-productive methods • teach shortcuts • aid customizing and modification
Help When Things Go Wrong	• identify likely problems/solutions • identify problems needing expert help • enable user independence from developers

As Exhibit 1.1 shows, user manuals should not only help the users get started, but stay apace of their evolving interests over time, ultimately reducing the users' dependence on the developers. It also shows that much of what used to be in manuals might be better situated elsewhere: online Help and reference screens, video training programs, reference cards, tutorial programs on disk, and so forth.

But even when the pages of a manual are on a screen, instead of in a binder, most of the criteria for good documentation remain unchanged. And, just as important, so do the dangers and difficulties.

1.2   Three Misconceptions About User Documentation

User manuals are *not* a form of literature. User manuals are *not just* reference manuals. And user documentation rarely consists of *one* book.

Even the noblest of intentions may be frustrated when certain misconceptions about documentation get in the way.

First, user documentation is *not* a form of literature. Writing user manuals is not just a matter of hiring a wordsmith, often a person with a recent degree in English literature, to beautify the language in the technical documentation. Although any scientific or technical enterprise can benefit from the contributions of people with backgrounds in literature and the humanities, although every company that produces user documents needs at least one professional writer or editor, and although every document should be reviewed at least once by a person who is expert in matters of language and style, it is nonetheless unwise to confuse literary and language skills with documentation skills.

Underlying this misconception is the dangerous notion that the only important difference between technical documents and user documents is that the latter need to be simplified and prettified for the benefit of a reader who cannot understand the technical documentation in its pristine, unedited state.

The truth is that, even if every sentence in a technical document were revised and restyled for clarity and correctness of language, it still might not serve the needs the user. Simplification is not enough. Removal of technical jargon is not enough. Literary or stylistic excellence is not enough—if, through lack of analysis, the text contains the wrong sentences about the wrong topics in the wrong sequence.

Second, user documentation is *not just* reference material. Reference material is *part* of user documentation: the part we use *after* we know how to operate the system. Reference material is the set of things we look up *after* we know what we need to know.

Not surprisingly, reference material is all that programmers and developers need to operate their *own* systems. They know how the system works, what it does, and when to use its features. If they need user documentation at all, it's only to "look up" some term or rule or value they have forgotten—or never bothered to memorize in the first place. Similarly, users who have general experience with a certain class of system may be able to implement a new system merely by looking up what is different about this particular one.

But what of the new users? If we provide them with complete reference materials, will they be able to operate the system? Will they know, for example, that every disk has to be formatted if the only place this is mentioned is in the reference manual description of the FORMAT command? And can we ever expect them to look up FORMAT when they have no reason to?

Often, when we entrust the writing of user documentation to technical experts, *reference* material is what they write. Even though it is true that "everything the user needs to know is in the book," it is also true that the presentation obscures the information so that it is almost useless.

Third, user documentation is rarely a matter of producing just *one* book. Although there are circumstances and systems in which all the users' information needs can be addressed with a single manual – usually a large and intimidating one – most systems need a library of documents and other information products. People who prepare user documentation should *presume* that there will be more than one publication, changing their minds only if they are convinced that one will be enough. Indeed, because user documents almost always come in sets or libraries, it is extremely difficult to design or write one publication well without knowing the associated documents that will also be available to the users.

Except for those increasingly available software products that are so English-like in their prompts, messages, and help screens, a single, encyclopedic user guide rarely serves the community of users – especially if there is a variety of users and especially if the system or software has multiple uses.

And this last point bears emphasis: some of today's software and systems are so versatile and powerful that they have literally thousands of potential uses. Of those thousands, only a few interest any given user. Does it make sense, then, to expect every user to deal with the same compendious manual?

Credit is due, of course, to those companies that enhance their manuals with complete tables of contents, indexes, and other search tools. Also to those who ensure that each page is uncluttered and readable. Still, putting all the information into one huge publication shifts the burden for using that information to the reader. The bigger and more complicated a manual, the more energy and skill are required of the readers. And my main objection to this one-book-fits-everyone approach is that I believe the burden for communicating useful information to the users is at least partly the writer's!

Do not misunderstand. I do not claim that a manual can be written in such a way that every reader finds it easy to use. Nor have I overlooked the fact that some users and operators will not read manuals, no matter how they are written. But I still assert that the responsibility of making a manual work – making it *usable* – rests with the designer and writer of the publication.

1.3 Why So Many Good People Write Such Bad Manuals

Many of the firms that should write user manuals write none. Most of the firms that write user manuals do not write enough of them or keep them up to date. And a substantial portion of the manuals written—even by the most sophisticated firms—are ineffective: clumsy, inaccessible, inaccurate.

There is a growing group of firms that consistently produce high-quality, readable user documentation. And a few more firms that produce it much of the time. Together, though, they are still a handful.

The typical case, however, is **no user documentation at all.** As I travel the country, I am still surprised at how many computer companies, engineering firms, software consultants, banks, and manufacturers have no user manuals or operating instructions for their systems or products. (Even more terrifying is how many have **no technical or system documentation** either.)

Those that finally succumb to pressure and try to write documentation are likely to produce unsatisfactory results: books that run the gamut from hastily-typed-and-unusable to expensively-typeset-and-unusable.

Why? How is it that companies smart enough to design an "automated teller" or a CAD/CAM system or a network that allows computers to talk to copy machines . . . how is it that these bright, resourceful organizations cannot manage to write an intelligible user manual?

There are two main explanations: first, some don't care; second, some don't know how.

Long before there were user manuals for computers, there were instruction books and assembly guides for equipment. And for as long as there has been such literature, much of it has been unreadable. Why? Because, traditionally, engineers and manufacturers do not like to spend time or money on these documents. The old joke is that the instructions are written by whichever engineer is not busy that day; but it is no joke.

Hardly anyone I know likes to write. A good many of the engineers, scientists, and systems analysts I know hate to write. And the writing they hate the most is explaining complicated, technical ideas to people who know less than they do.

That many firms are indifferent to user documentation is apparent. They set aside almost no time to get it written and often assign it to people with other "more urgent" things to do. Or they delegate it to a junior employee, who lacks the authority and leverage to do it well. Furthermore, they spend as little as possible on production and artwork; most hire no artists or technical writers to help. Indeed, the typical user manual written in America is put into use without ever having been reviewed by a skilled editor.

In those firms that do care, matters are a little better. Still, though, the central problem affecting the writers of user manuals – including professional technical writers – is that they have not received enough guidance and instruction on how to write them. Most people about to write a manual have never written one before; only a few have a "good one" to refer to as a model. Indeed, most of the newly hired documentators in America are working for supervisors and bosses who have never written a manual.

Even though there is about thirty years' worth of research on techniques that make documents more accessible and readable, most people, including more than a few professional technical writers, have read none of it. What constitutes good writing is still "art," in the least favorable connotation of the word: hunches, intuitions, preferences, instincts. Too many discussions about user manuals – especially about editing and refining them – devolve into disputes about personal preference.

So, in the extreme, stereotypical cases, user manuals are often written either by technical experts who dislike the job, give it as little effort as possible, and use no formal criteria to decide if the job was done well. Or, at the other extreme, by artisan technical writers, who bring all their intuitive and stylistic sense to the project, but who lack the theories and formal criteria needed to decide if the job has been done well or to justify its cost to skeptics.

The purpose of this book is to convince both groups, and everyone else involved in user documentation, that *usability* (how appropriate, accessible, and reliable a manual is) can be defined and measured objectively. And, furthermore, that it demands participation of both groups. **Analysts and technical experts cannot, working alone, produce usable user manuals.** Not because they write badly; they do not write worse than people in other learned professions. Rather, because they know too much and, with few exceptions, cannot make themself clear to less knowledgeable readers. And neither can most technical writers, who, in the typical company, must beg for "input" from the technical staff. (Later, the intended user will also get into the process.)

Good user manuals also need a change of perspective. Manuals – effective, workable manuals – are not so much written as *designed*. (The same statement is true for computer programs.) They are *engineered*. The key is to think of them as *devices*.

1.4   A New Approach: Manuals as Devices

As long as writers are thought of as artists and user manuals as works of art, then neither professional writers nor analysts/engineers are likely to produce usable manuals. The key is to think of books as *devices*. This notion changes several related attitudes, such as: how books should be developed; what can be expected from the reader; what criteria should be used to judge the book; and how the cost of the book should be justified.

If manuals or other information products are thought of as works of art, it will be extremely difficult to change the methods people use to develop them. If instead each book or videotape or series of screens is thought of as a device, with a set of functions, then *usability* becomes possible.

It is impossible to escape the analogy between books and computer programs. Manuals affect readers the way programs affect computer hardware – except that readers are far more fallible and have far less reliable memories. Manuals pass instructions and data to their readers, who then operate the system correctly and productively.

This is not just calling manuals by another name. To think of books as devices is a serious change in perspective for many writers and analysts. To do so obliges us to rethink our notions of user documentation, to change a whole cluster of related attitudes.

EXHIBIT 1.4   Manuals as Devices

	"Work of Art"	"Device"
Greatest Effort	Composing the draft	Testing, refining the design
Procedure	Compose and polish	Specify-Model-Test
View of Reader	Independent, resourceful	Dependent, error-prone
Basic Criteria	Style, appeal, preference	Meeting the specs, usefulness
Advanced Criteria	Beauty, elegance, "class"	Maintainability, reliability
Cost Justification	Unpleasant necessity	Productivity, efficiency
Cost Objective	Spend little as possible	Increase profit and ROI

As Exhibit 1.4 shows, the first change is in our conception of the *writing process*. If writing manuals is an art, then the creativity is in the drafting, the composing of the words and sentences; writers who think of themselves as artists spend most of their time writing and polishing the draft. In contrast, if a manual is a device, then the creativity is in the *engineering*, writing the specifications, building and testing models – all of which precedes the execution of the design (the draft).

Also, our *view of the reader* changes. The artist views readers as independent and active; the burden is on the readers to find things and apply them correctly. If books are devices, though, readers are less independent. Instead, they rely on the design of the book; the burden shifts to the documentor. As I shall explain later, in this view the writer controls the attention of the reader, much as software controls hardware – and for similar reasons.

The *basic criteria* for judging publications also change. If a book is art, then the basic criteria are style and appeal – a sense of correctness and craft, peculiarly understandable to the writer but difficult to explain to others. If the book is a device, though, the basic criterion is whether it meets the specifications, performs the job it was designed for.

The *advanced criteria* also change. For artists, a very good book is one that has beauty and elegance, slickness, "class." If a book is a device, though, the advanced criteria are from engineering: maintainability (how easy it is to update and enhance the book) and reliability (how often the book "fails" in use).

And finally, the *cost justifications* are entirely different. The hardest task for the artist-documentor is to justify the cost of user documentation. Beyond convincing management that at least some user documentation is an unavoidable business necessity, the artist is usually powerless to justify expensive processes and products. "Class" and "style" are not usually persuasive. In contrast, the justification for books as devices is that they save or make money. The documentor who thinks about books as devices should have little trouble justifying the expense; each device should return much more than it costs.

1.5   The Four Functions of User Publications

Traditionally, user documentation is divided into two large categories: *instruction* and *reference*. Practical experience shows, however, that instruction should be further divided into two large subcategories: *tutorials* (teaching aimed at the novice) and *demonstrations* (teaching aimed at the more-experienced user). Still further examination reveals a fourth category: *motivation*—writing aimed at getting users to do something they are reluctant to do.

Every unit or module in a well-made user publication will perform exactly one function. But what are these functions? What does user documentation do? Traditionally, user documentation has been expected to help in two ways:

- *Instruction*, teaching people how to run or operate the system or product
- *Reference*, giving people key definitions, facts, and codes that they could not be expected to memorize

But the change in the world of systems and their users suggests the need for a more refined breakdown. From my consulting experience, I conclude that *instruction* is too large to be considered one category. Instead, I propose that instruction be broken into *tutorials* and *demonstrations*. Tutorials are those instructional materials intended to train or orient neophyte users; demonstrations are those materials aimed at teaching a process or activity to a competent or experienced reader.

Tutorial documentation, which nowadays is often written on diskettes rather than in books, is the newest form of user documentation, and the form that gives most trouble to traditional technical writers and, especially, to the programmers and managers who have been conscripted into the job of writing it. Further complicating matters is the rising prominence of a reader I think of as Reader X, a person who is intimidated by books and has seldom been able to learn successfully from reading.

Demonstration, as the name suggests, is teaching by showing. Aimed at a person who knows generally what to do with the system, demonstration shows whole processes or activities, from the top down. In contrast, tutorial documentation ordinarily begins from the bottom, with elemental definitions and concepts.

Reference documentation—what some programmers mistakenly equate with user documentation—is a compressed presentation of facts and information, useful only to people who know what they need to know.

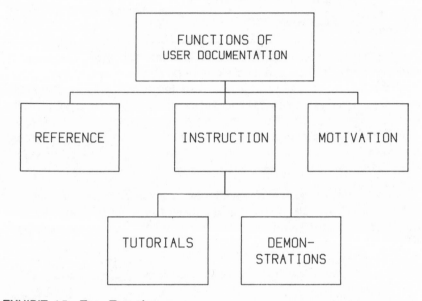

EXHIBIT 1.5   Four Functions

Highly experienced operators and users need nothing else; new and intermediate operators and users need much more.

The change in the community of users has created the need for a fourth function: *motivation*, documentation designed to get people to do what they are reluctant to do. In effect, motivation is the selling of ideas and methods. And, although not every user manual needs it, far more need it than have it. Put simply, many system problems can be blamed on reluctance, not ignorance. Whether from fear, greed, or just laziness, many operators and users simply will not use systems the way we think they should.

Obviously, motivation calls for a different kind of writing from that appropriate for teaching — a different form of control. Obviously, trying to teach a neophyte clerk calls for a different technique from that used to teach an experienced installer.

2

Needs: How User Manuals Fail, Succeed, or Prosper

2.1   The Main Goal of User Documentation: Control

Here is a paradox of effective writing: To communicate well one must respect the independence and intelligence of the readers, but must not rely on them. Most readers resent being spoon-fed or instructed, but many of them need just that. For user and operations manuals, the best strategy for writers is to adapt to the weaknesses in typical readers, to assume *control* of the communication.

If a user manual is regarded as nothing more than a packaged collection of facts, a binderful of miscellaneous information, then its usefulness depends primarily on the skill and resourcefulness of the reader. In contrast, if a manual has been engineered to suit the interests and ability of the reader, then the user is to some extent prevented from misusing the material.

When I use the word *control* to describe this approach, many critics object. Any talk of controlling people elicits sincere ethical objections.

I do not mean to suggest, however, that writers should coerce readers; nor do I think that readers can be compelled to understand a bit of text or diagram. Rather, I am proposing that anyone who wants to write effective user manuals will have to learn to regard the readers as complicated information processing systems and to **control for the sources of noise and error in such systems.**

Even though we still know relatively little about how people read or understand, we are not completely ignorant. We know many things that cause readers to get lost and make mistakes; we know something about the effects of memory. At the very least, the documentors must remove from the manual anything they *know* will interfere with its correct use – especially those things that almost invite the reader to misuse it.

In effect, my claim is that manuals and other information products affect readers in much the way that computer programs affect computers: they *control* their operations. And just as an underdesigned computer program will cause the system to break down, or interrupt, or consume too many expensive resources, so will an underdesigned manual cause readers to get lost, make errors, even shut down their work. Just as an undertested program will throw off indecipherable bugs nearly every time it is used, so will an undertested manual generate mistakes and inconsistencies nearly every time a reader consults it.

The objective is control of the readers/users, for their own advantage. The aim is to devise manuals that compel readers to find what they need, in the most efficient sequence, and with a level of effort that does not either discourage them or lower their productivity.

(I do not recommend this view for all writing, or even all business writing. Literature depends on the imagination, experience, and intellectual talent of the reader for its effectiveness. Excellent essays often demand close reading and study. But user manuals that must be studied to be understood are, in general, ineffective.)

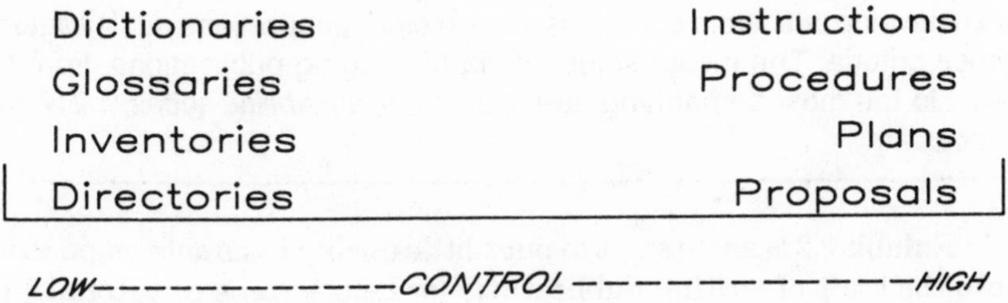

Dictionaries	Instructions
Glossaries	Procedures
Inventories	Plans
Directories	Proposals

LOW — — — — — — — — — — *CONTROL* — — — — — — — — — *HIGH*

EXHIBIT 2.1   Continuum of Control in Publications

Every user publication fits somewhere on the continuum that appears in Exhibit 2.1. **At the highest level of control are those publications meant to be read from the first word to the last, without omissions, without skipping or skimming.** Most notable in this group are installation plans, assembly instructions, introductory training programs, new product proposals and specifications. In this category is nearly every document that is *incremental* (presenting an accumulation of increasingly complicated facts), *procedural* (presenting a set of steps or activities that constrain one another), or *argumentative* (presenting an inductive or deductive chain of assertions).

At the other extreme are publications that *no one* would ever read in sequence: dictionaries, glossaries, inventories, and directories — alphabetical and numerical listings of reference material. And yet, even at this end of the continuum, there is still a benefit in controlling the reader. A well-designed reference directory allows the user to find information quickly, with "one pass," to complete the search without needing to skip and detour, and, finally, to exit promptly with the needed information.

2.2   Four Criteria for Effective User Documents

One of the main steps in converting the creation of user documentation from an intuitive craft to a kind of document engineering is to define formal criteria that can be used to test the desirability of alternative manuals (or designs for manuals). If the organization can agree on the criteria, it can then develop indicators or metrics that assess a publication's quality against those criteria. The most useful criteria for judging publications, from the least to the most demanding, are *availability*, *suitability*, *accessibility*, and *readability*.

Exhibit 2.2 is an attempt to put a little quality assurance engineering into the craft of writing publications (or help screens or tutorials) for users. Clearly, there are at least four levels of increasingly higher standards of quality, starting with *availability* (Is there anything at all?) and moving through *readability* (Is it in clear, easy-to-understand English?).

Availability

There are still developers who provide *no* user documentation. Generally these are DP shops or companies in which everyone is a programmer. Such organizations simply are not *aware of users* – what they do, what

STANDARD	REQUIREMENT
READABILITY (Fluent, Simple-to-Read)	PROFESS. EDITING
ACCESSIBILITY (Comfortably organized)	DESIGN
SUITABILITY (Aligned with tasks/interests)	ANALYSIS
AVAILABILITY (Present when needed.)	USER– AWARENESS

EXHIBIT 2.2   Criteria and Levels of Quality

they know, how they work. And until the unit hires someone with such an awareness, it will continue to overlook user documentation.

Suitability

Today, most vendors and developers provide at least some user documents. Sadly, though, they tend to subscribe to the encyclopedic view of the user manual: Put everything in one big volume and let the users fend for themselves. What they ought to do is *analyze* what documents are needed: align particular publications with the tasks and interests of particular readers. Until that happens, their "comprehensive" documents will often be unsuitable, unusable, and unreliable.

Accessibility

It is possible for a book to contain exactly what the user needs, but still be organized in a useless tangle. As a result, readers have to skip, branch, loop, and detour from page to page—until they get lost. In software engineering terms, the book, because of its excessive number of GOTOs, is unreliable. That is, the paths through the book are so complicated that even a skillful reader will probably get lost. Only firms that *design* their books for accessibility (and test and debug the designs) produce smooth-reading, GOTO-less user manuals. User manuals that are both suitable and accessible are likely to be called *task-oriented*. That means the developer has analyzed what the users do, how they use the system and product, and what information they need. The task-oriented manual, then, is one that contains only what the user needs, in the most useful sequence.

Readability

Even when a book is suitable and accessible, its ultimate quality resides in its readability—how easily and accurately it can be understood by its intended audience of users. Still, too many DP professionals regard matters of language and style as "frills"; hundreds of manuals and instruction books are published without so much as a cursory review by a professional writer or editor. Only *professional editing* can produce manuals of the highest quality.

Users, then, want manuals of a quality that can be attained only with *user-awareness*, *analysis* of information needs, *design* and testing, and *professional editing*. The thoroughness and extent of these are, of course, subject to constraints. Most deadlines must be met; certain opportune moments must be exploited.

2.3   Three Classes of Errors

For user documents to score high on the four criteria of quality, they must have been planned and designed the right way. If documents score low on one or more of the criteria, the failure can be blamed on one or more classes of errors: *strategic*, *structural*, or *tactical*.

A point rarely appreciated is that **much of what is wrong with user documentation is the result of mistakes made before the draft was written, and that the most serious flaws in user publications are nearly impossible to correct after the first complete version of the publication is drafted.**

There are three broad classes of errors that can undermine user documentation – and only one of them can be corrected in the editing stage. The first class, *strategic errors*, includes failures of planning and analysis: failure to define exactly what documents were needed to serve exactly what audiences in the completion of exactly what sets of tasks. These are the main strategic errors:

- overlooking the need to plan or analyze documentation requirements
- allowing the product or system to shape the documentation, instead of the users' interests and tasks
- assuming that only one encyclopedic manual is needed
- refusing to adapt to the vocabulary and reading skills of the intended audience

Structural errors are failures of design and modeling: insufficient outlining, lack of rigorous review of the outlines, failure to test the plan of the publication before writing a detailed draft. Even if the planners have made no strategic errors, structural errors can still lower the suita-bility of the manuals and, more relevant, so reduce their accessibility as to make them unusable. These are the most common structural errors:

- using little or no outlining or other document specifications
- relying on superficial, "grade school" outlining methods
- failing to submit outlines and specifications to harsh reviews (walk-
throughs)
- omitting the intended users and readers from the design process

Tactical errors are failures of editing and revision: inconsistent nomenclature, mechanical errors of grammar and spelling, clumsy "first draft" style, ambiguous sentences. Tactical errors occur either when no one in the organization knows enough about clear communication to edit a first draft or when the company simply does not give enough time for the editors to do what needs to be done.

Notice the paradox. On the one hand, it is a serious mistake to publish a manual that has never been reviewed by a competent wordsmith. But, on the other hand, it is even more dangerous to believe that the skills of a wordsmith can compensate for having written the wrong publication.

In effect, then, a usable manual must pass three tests:

- The *strategic test* proves that the manual is well defined, aligned with a specific audience and use, part of a coherent set or list of information products.
- The *structural test* assures that the elements in the publication are in the most accessible, reliable sequence.
- The *tactical test* assures not only that the sentences and exhibits will be free from distracting errors and clumsy passages, but also that the reader will be spared the risk of finding out what ambiguous passages mean the hard way.

EXHIBIT 2.3   Three Classes of Error

Error	Source
STRATEGIC Failures of Planning/Analysis	• poor definition of audiences • poor definition of tasks • lack of overall plan
STRUCTURAL Failures of Design/Modeling	• lack of substantive outlines • poor tests of outlines • omitting users from the review
TACTICAL Failures of Editing/Revision	• careless inconsistencies • "first draft" style • subprofessional editing

2.4   The Enemies of Quality Documentation

Why do many organizations persist in producing low-quality, error-filled manuals, even after they have been shown alternatives? Because there are powerful organizational and psychological forces that work against high quality: *fear of conflict*, *myopia*, *impatience*, and *perfectionism*.

Many companies and many writers heartily endorse the methods in this book – but nevertheless fail to implement them. They concur in the definition of the problem; they approve the criteria; they also agree that the methods presented later in this book can solve the problems and produce high-quality documents. But they still fail to implement. (Software quality consultants have reported a similar problem.) Why?

When people fail to do what is clearly in their best interest, usually it means that there are other, competing interests at stake. In this case, what prevents many people from developing high-quality documentation, and what also often prevents them from developing high-quality systems in general, is a set of related *enemies*, forces that draw people away from the path of quality and onto the path that might best be called "quick-and-dirty." The enemies are shown in Exhibit 2.4.

EXHIBIT 2.4   The Enemies of Quality Documentation

Enemy	Tendency	Consequence
Fear of Conflict	Avoiding conflicts and clashes	Delayed testing and problem-solving
Myopia	Preoccupation with short-term costs	Loss of long-term benefits
Impatience	Worship of arbitrary deadlines	Underdeveloped products/ publications
Perfectionism	Avoiding estimation and prediction	Indecisiveness, useless delay

Fear of conflict is avoidance of all confrontations over matters of policy and priority – especially those involving a clash between divisions or departments. Similarly, it is an unwillingness to communicate with people who are hard to understand or who ask too many questions or who raise objections. In most large bureaucracies – business or government – avoidance of conflict seems to be a more influential goal than even the quest for profit or effectiveness or quality.

Myopia is an inability to consider any costs or consequences other than those that can be seen or counted right now, in the current planning or budgeting period. It is an unwillingness to evaluate downstream costs and benefits, a stubborn refusal to weigh short-term costs against long-term benefits. Most publications departments are oppressed by such concerns and are therefore prevented from doing things that would save the company much more than they cost.

Impatience is a consuming preoccupation with meeting deadlines, many (or most) of which are arbitrary. It is an attitude related to myopia, but with an emphasis on haste and on superficial measures of output (number of pages, number of lines of code . . .). Complicating matters further, many managers give lip service to quality but, in fact, enforce their own impatience instead, rewarding timely "quick-and-dirty" work and punishing anything that is late.

Perfectionism is a discomfort with planning and estimation, an unwillingness to make predictions that might be wrong, or to discuss incomplete or unfinished work. Often, perfectionism is a technically respectable form of indecisiveness. A reluctance to make commitments and judgments with incomplete data is really a reluctance to make any decision. Frequently, perfectionists are more confused than aided by additional data.

In the presence of these barriers, the most ardent documentors will surely be frustrated. They will be told: "Worry about this problem later." (fear of conflict); or "Color costs too much." (myopia); or "Whatever you have done by Friday will have to be good enough." (impatience); or "We'll decide what kind of graphics to use after we've seen the first draft." (perfectionism).

3

"Usability": Documentation as a System

3.1   From "Idiot-Proof" to "Usable"

When engineers and inventors devise truly new products or techniques, they frequently worry least about whether the product is easy to use. In the 1980s, though, *usability* has become one of the main objectives for designers of computer products.

Today's computer systems, for the most part, do the same tasks that the computers of the 1950s performed – except faster, cheaper, and with less human effort. Since the advent of the data processing industry, then, there has been an evolution of criteria; with each era, the ante has been raised.

As Exhibit 3.1 shows, the prevailing criterion of system quality in the 1950s was *whether the system worked at all*. Gradually, analysts and engineers shifted their attention to the economies of *efficiency* – throughput and cycle times, resources used, and so forth. When we had to pay $60 per second to lease a relatively slow central processing unit, efficiency was of the essence.

As machines and memory have dropped in price, though, the emphasis on efficiency has decreased in many places. Nowadays, it often costs more in personnel expenses to make a machine efficient than could be saved in the efficiencies. Today, the most important, most frequently discussed technical criterion is *maintainability*, the ease with which a system can be fixed, adjusted, or enhanced.

RELIABILITY/USABILITY (Lack of Interrupts)	1980's
MAINTAINABILITY/MODIFIABILITY (Ease of service/enhancement)	1970's
EFFICIENCY (Uses machine resources well)	1960's
ANY PERFORMANCE (Does it run at all?)	1950's

EXHIBIT 3.1   Evolving Standards of System Quality

Moving into the 1980s, the theme has changed somewhat. Although many DP and MIS organizations have still not entered the 1970s – that is, they are still concocting unmaintainable systems without benefit of the new development methods – the latest chant is "user-friendliness," making the system easy to use.

Computer technology, then, has completed an entire cycle of development: for the most part, it still does the same things it did in the beginning – but in a much friendlier manner. The typical operator of today's computer is not a mathematician or programmer, but rather a clerk or business person, or even a 10-year-old child. Engineers may not use such terms as "idiot-proof" to describe the new systems, for to do so hearkens back to an earlier epoch of computer technology, when the user was presumed to be an expert.

Usability – a more sensible term than "user-friendliness" – is an *engineered constraint*. That is, the built-in characteristics of a device, system, or program put an upper limit on how easy that entity will be to use. A task that calls for 20 keystrokes will be harder than a task that calls for 2 – no matter how the instructions are written. A "Clear Display" button right next to an "Insert Character" button is more likely to produce an inadvertent clearing of the display than a button several millimeters removed, despite the warning in the manuals.

There are competing notions of usability, of course. For example, making a system easier to learn at first is not always consistent with making it easy in long-term every day use. Usability is a consequence of how well the system has been defined, specified, and tested. It comes from doing the analysis and design well – not from writing heroic user documentation after the fact.

For documentors, moreover, the term *usability* has two important, related meanings. First, it refers to the ease with which a system can be operated; second, it refers to the ease with which the documentation can be operated. Put another way, if the user documentation is also regarded as a system – the position I urge throughout – then it follows that the **usability of the documents restricts the usability of the computer system.** When the user documentation is extremely usable, then the computer system will be no harder to use than it must be. If the set of user manuals and other information products is the best possible, of high usability, then the system documented will be as easy to use as its engineering permits. But every time there is an error or shortcoming in the user documentation, the system loses usability.

3.2   The First Law of User Documentation

Each system or product has an inherent usability; each document associated with the system has its own usability. But even the best documents cannot compensate effectively for flaws in the system itself. The first law is: Clean Documentation Cannot Improve Messy Systems.

Once a system has been implemented, there is little anyone can do to change its overall usability. Although it can be improved, the improvements are likely to be superficial. One especially ineffective way to increase the usability of an existing system is to try to cover its flaws with "especially good" user documentation. (In effect, we leave the hole in the road but post warning signs; mistakenly, we then believe that the hole is no longer a danger.)

But we cannot turn a jungle into a garden by drawing a map of it. Nor can we make an intimidating and complicated procedure user-friendly by writing a slick operations guide. Whenever user documentation is written after-the-fact (first you develop, then you document), it cannot compensate for failures of analysis, design, or coding.

Although it may be odd to say so in a book about user documentation, it is wrong to expect user manuals to do too much. We should not expect them to ameliorate engineering and programming mistakes.

We must never forget that a simple procedure, explained well, is clearly simple. A difficult procedure, explained well, is still difficult. A dangerous and trouble-prone procedure, explained well, is clearly dangerous and trouble-prone. Just because bad writing makes procedures harder to follow, it does *not* then follow that good writing will make them easier to follow.

The only way for user documentation to improve a system is for it to be written before the fact. Writers of users manuals, as the "first users" of the system, can discover ways to improve the system that developers are unlikely to see. And, if they write the documents clearly enough, before the system is etched in disk, there may be time to redesign the system.

As Exhibit 3.2 shows, if user manuals are written (or at least designed) during the functional specification of the system, they can be used as engineering models; developers can detect and correct errors and unreliabilities in the human part of the system—the so-called "user interface." Even during the design stage, there is still a chance that the discovery of hard-to-explain procedures can be reflected in improvements within the modules of the system being documented; it is still practicable to make these changes. At the trailing end of development, however, the writer of the manuals is more-or-less stuck with the system as it is.

EXHIBIT 3.2 The First Law of User Documentation

If user manuals are developed during . . .	Then, the manuals can . . .
Functional Definition of the Product/System	• Clarify procedures and policies • Identify unfriendly elements • Increase chances for user satisfaction
Product Design/Coding	• Clarify bugs and errors • Identify causes of unreliability • Force designers to make early decisions
Distribution and Use	• Help users adapt to the product • Warn against bugs in the system • Disclaim liability

In sum, any procedure that defies simple instruction by a competent writer will probably be hard to learn. Any activity that cannot be explained without a railroad timetableful of exceptions and cautions probably cannot be used efficiently. Under the right circumstances, these problems become known before the system or procedure is locked-in, before the prospect of changing becomes too costly and unpopular. The documentor helps change the procedure; sometimes, the change eliminates the need for several pages of tortuous discussion. Nothing is harder to document than a clumsy, unreliable procedure. And vice versa.

Note the irony: the better the design of a system, the less user documentation it needs. Documentors who discover flaws in the systems soon enough can reduce the quantity of writing they must do.

3.3 Defining and Measuring the Usability of Publications

If the objective is to design and engineer publications for usability, and if the process is to be more than "artistic," then there must be formal criteria and measures. A proposed Index of Usability: the more often the intended reader must skip material or reverse directions while reading, the less usable the publication.

Even though good documentation is no substitute for good systems, usable publications are an essential ingredient in successful implementation.

Moreover, it should be possible to define some measure – an Index of Usability – in such a way that it can be applied **before the manual is written**, in time to correct whatever flaws and bugs it discloses. Remember the essential point: The later in the life of a product, the more expensive to change it, and, therefore, the less likely it will be changed.

From my experience, I conclude that the most powerful Index of Usability is **the number of times the intended reader must skip material or reverse directions to use the publication.** (Obviously, this is an *inverse* predictor.)

I do not mean to suggest that all user publications should be written so that every user reads them from the first word to the last. In fact, relatively few publications will be like that. I do mean to say, however, that any skip or loop in the manual – intended or not – lowers its usability.

Note that the definition of the Index includes the phrase "intended reader." Clearly, readers with different interests and backgrounds would use the same manual differently; one would read every word in sequence while another would skip and glean. Indeed, the same reader, after one or two one-directional passes through a manual, would later skip and glean. Obviously, the more diverse the audience for a certain manual, the harder to make it usable for everyone.

Interestingly, the skips and loops (branches, detours, and GOTOs) can be grouped into three classes, corresponding to the three main errors of documentation:

Strategic errors cause the largest skips and spins. Failure to align the books with the readers will send the readers jumping from book to book, until they finally find what they need – or give up. When a user needs two books to do one job, then the selection and partitioning of the books does not reflect the needs and interests of that user. If a user must ignore 98% of a publication, then it must have been designed for someone else.

EXHIBIT 3.3 Measuring Usability

Criterion	Error	Consequence
AVAILABILITY	Strategic	• jumping from book to book • needing two books for one task • needing to ignore most of the pages
SUITABILITY	Structural	• jumping from front to back • never reading pages in sequence • searching for exhibits, tables . . .
ACCESSIBILITY	Tactical	• stopping to notice mechanical errors • getting stuck on inconsistent terminology • rereading difficult passages
READABILITY		

Structural errors cause medium-sized loops and skips. The manual calls for frequent jumping from front to back, especially when the text refers to charts, tables, and exhibits that are elsewhere. (NOTE: **The greatest single barrier to the usability of a manual is the separation of text from the exhibits referred to in the text.**)

Tactical errors cause the smallest GOTOs, usually within a paragraph or page. Because the editing is poor, the reader must loop on unclear sentences, inconsistent nomenclature, distracting errors of grammar, and so forth. Although these are the smallest breaches of usability, they can be powerful enough to undermine even the right book with the right structure.

Defined in this way, usability can be assessed during the earliest stages of planning and design. Strategic errors will produce loops in the publication plan – if we test the plan. Structural errors will produce loops and GOTOs in the outline – if we build and test the outline properly. Only tactical errors will wait until the draft is completed.

3.4   Usable Manuals Are Vertical, Not Horizontal

Horizontal manuals are those that describe everything that *could* be done with the product or system and that are organized according to the characteristics of the object described. Vertical manuals show how to do specific things and are organized according to the procedures or tasks to be carried out by the reader. Horizontal manuals are *reference books*; they are generally unusable as teaching or demonstration documents.

In the early days of computing, programs did one or two things; there were few alternatives or "user-defined options." So, the Run-Books for these programs were straightforward, linear, easy to follow, task-oriented.

Yet, the trend today is toward products that perform as many applications as you can think of: high-level languages; data base management systems; packages that can be used to perform nearly any statistical analysis or generate nearly any of the common graphic displays.

Documentors should be aware that there is a potentially significant strategic problem in writing the user and operator materials for these multipurpose systems. Many users are unresponsive to the discussion of dozens of generic skills and features. For example, a doctor or warehouse manager may be uninterested in "How to Define a Variable," even though, from the developer's point of view, this is one of the most important things to learn.

Here is the paradox: Even the most versatile software products — whether they are the data base managers that run on giant mainframe computers or the various competing "spreadsheet" products designed for the popular personal computers — are used in **applications**. Although the people or company who invented the product may be terribly proud of its versatility, and even though some sophisticated users, mainly experienced computing hands, can think of a hundred useful things to do with the product, most users want to learn how to do *their* projects, solve *their* problems, improve *their* performance.

There are three important kinds of information, then, that must be in the manuals:

- instructions for installing the product in the *user's* place of work

- instruction in how to use the product to create *user-invented* applications, with examples drawn from the *user's actual needs* and interests, or

- *application programs*, already created to solve problems for the *particular user*.

Of these three types of material, the first two seem far more reasonable to the contemporary developer than the third. The third item seems unprogressive, though. Why would anyone want to write a user manual that treats an amazingly versatile software product as though it were a single-purpose application? Does not this devalue the product?

No, quite the contrary. It makes the benefits of the product more apparent to the user and, thus, makes the manual more usable. Readers must be given the right to have their own interests! They may choose to use their versatile software/hardware configuration as though it were nothing more than an appliance: like a phonograph or a video game.

Consider the pair of outlines in Exhibit 3.4. Do you notice that they cover many of the same topics? Although there are some interesting differences in style (to be discussed later), the main difference is that version A is horizontal; and version B is vertical. For the planner to use version A, he or she will have to skip and loop incessantly, and this lack of usability in the manual will detract from the usability of the product. Version B is task-oriented. In use, version B will be much more reliable.

EXHIBIT 3.4   Horizontal versus Vertical Organization

Version A (horizontal)	Version B (vertical)
1. System Administration	1. Installing Your System
1.1 Defaulting security features	1.1 Backing-up distribution disks
1.2 Defining configuration	1.2 Defining your company's security rules
1.3 Initializing files	2. Creating the Files
2. File Management	2.1 Explaining your chart of accounts to the system
2.1 Defining a file	2.2 Entering your current books
2.2 Reading files	2.3 Writing a set of "planning factors" programs
2.3 Linking files	3. Applications
2.4 Updating/maintaining files	3.1 How to analyze profit and loss by cost center
3. Input Preparation	3.2 How to compare year-to-year cost differentials
3.1 Worksheets	3.3 How to forecast revenues and costs
3.2 Data entry	3.4 How to simulate alternative budgets
3.3 Data editing	3.5 How to simulate earnings performance
4. Outputs	4. Presentations
4.1 Printing	4.1 How to make TREND charts
4.2 Graphics printing/plotting	4.2 How to make SHARE charts
4.3 Storage	4.3 How to make COMPARISON charts
Appendix I Alternative System Configurations	4.4 How to make TABLES
Appendix II Sample Outputs	
Appendix III Error Messages	

3.5   Controversy: Usability Versus Economy

To do many of the things that make a user manual more usable leads to repetition, duplication, even what some publication managers would call waste. Often, the objectives of usability and economy are in conflict, and the conflict must be resolved through policy and negotiation.

Many of the people who lead the user documentation activity in the their companies are publications managers. And, ironically, the top priority for a publication manager is often *production economy*, the need to keep the above-the-line costs for manuals and publications as small as possible.

Sometimes, though, the downstream costs and diseconomies associated with mediocre user documentation overshadow short-term savings. Many of the practices that reduce the production, distribution, and storage costs of manuals, moreover, also reduce their readability. For example: Have you ever met a reader who enjoyed working from microfilm or reading fine print? Do you know anyone who likes to flip back and forth between two sections of a book, taking comfort in the fact that the publisher was able to avoid duplication?

Consider the methods used to reduce the *area* of information and, thus, the number of pages. The combination of small print and narrow margins is the easiest way to reduce the number of pages – and the associated printing, mailing, and filing costs. One of the large national technical information services, for example, insists on single spacing and ¾-inch margins; the result is densely packed, nearly unreadable reports.

Similarly, many editors and publishers object to blank spaces, half-empty pages. But this book advocates (and practices in its own format) the policy of beginning each new section at the beginning of the next page; the result is loosely packed, more readable documents.

The most controversial issue is *redundancy* – a term that most of us have seen as a criticism of our reports and essays in school. Yet, redundancy is not always a term of criticism. In engineering, redundancy refers to the existence of deliberate "back-up" procedures and equipment, technology that allows the system to keep working even when the primary device fails or malfunctions. From the extra buttons sewn into a good suit to the three or four extra sources of power to drive the coolant pumps in a nuclear power generator, the idea is the same. Redundancy means reliability.

In communications engineering, redundancy compensates for the noise and entropy in a communication channel. The safest way to get an undistorted signal through a noisy channel is to send it more than once. That's why pilots repeat themselves when they talk to the flight controllers ("yes, affirmative") and that's why electronic funds transfers are sent at least twice, and then checked for parity.

In some ways, even large type, wide margins, and white spaces at the ends of sections are forms of typographic redundancy, allowing the channel to be less cluttered with information. More obvious is actual *repetition*, deliberate use of the same text and exhibits in more than one place – unthinkable to many publication managers. And, even more interesting is a practice severely discouraged by nearly every editor of technical journals: the use of art and diagrams that restate what is already in the text – saying in a graph what can be said equally well in a sentence or paragraph.

A fundamental belief among people who worry about the the short-term cost of technical publications is that artwork and illustrations should never be used unless they are necessary, unless they can show something that cannot be expressed in conventional sentences. Yet, I propose that well-made user publications will "back-up" their sentences with pictures (and vice versa). Why? Because there are word-readers and chart-readers, and if we want to adapt to the audience, and control the transaction, we have to write for both.

This is not to say that all deliberate repetition and redundancy is desirable or even effective. Some of it is inappropriate; some, irritating; some, even inconsistent with the goal of usability. Moreover, the amount and style of redundancy are a function of the publication and audience. An operations guide probably needs more than a textbook aimed at graduate students. And an operations guide aimed at inexperienced clerks or technicians needs more than one aimed at their experienced counterparts.

Although there will be an occasional, fortunate case in which economy and usability are compatible, most of what makes books more usable – including such items as durable, heavy paper stock and color printing – may seem expensive and wasteful at first. But only if we take a short-sighted view of costs. In the longer view, the benefits in efficiency and productivity can save thousands of times what they cost. And in the broader view, money spent for more-readable documentation eliminates or contains the costs of field service, hot-lines, training, troubleshooting, and a variety of other expensive services.

Good documentation should pay for itself many times over. And, consequently, the company with the best user manuals is the one in which the champion of quality user documentation is someone who worries about *all* the costs, not just this quarter's publications budget.

3.6   The Ultimate Tests: Reliability and Maintainability

The Usability Index (the degree to which manuals are free from skips, branches, loops, and detours) is not arbitrary and it is not esthetic. It bears directly on the cost of implementing and supporting systems. Usable manuals are more *reliable*, that is, less likely to fail in use. They are also more *maintainable*, that is, easier to fix or adjust when they do fail.

If documentation is a system, and if each manual is a component or device in that system, then each manual should be built for reliability and maintainability as well.

But what is the "reliability of a manual"? In what sense does a manual fail? Can we appraise a manual by tabulating its Mean Time Between Failures (MTBF being the most popular reliability metric in engineering)? Does a manual really break down?

A manual can be said to have failed if the user/operator is **unable to work because of it** – if a mistake, malfunction, or interruption can be blamed directly on the manual. Failures can result, then, from omitted information, incorrect information, ambiguous or contradictory information. Or, failures can result from an arrangement of materials that raises appreciably the effort needed to find information, leading to false starts, frustrated efforts, or "improvised" solutions to problems that cannot be handled with the manual.

It is difficult to say which failures are more costly: mistakes or interruptions. Interruptions are more common, since in many shops "downtime is offtime," and who doesn't enjoy a paid break. But, probably, the most expensive consequences are the destroyed files and other horrendous crashes caused by a "reasonable" misreading of the manual.

The first main justification, then, for the Index of Usability is that there is a direct correlation between the number and complexity of the skips and loops in a book and the number of errors and breakdowns likely to occur. The more paths through a publication, the higher the chances of taking a wrong path. The more discontinuous movement through a book, the higher the chances for a wrong move. The more choices for the reader to make, the higher the chances for a wrong choice.

This concept applies most directly to readers with limited experience, and especially to those with modest reading skills. But no one should mistake this principle as applying only to people who have trouble with complicated books. Even though some users are accustomed to tangled, unreliable books, no one likes them and no one is as safe trying to read a typically clumsy mess as a one-directional, GOTO-less manual.

Reliability can be treated as a target. Documentors who are trying to reach Reader X (my name for people who have trouble learning from

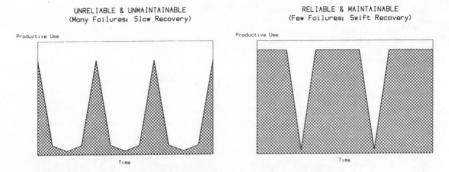

EXHIBIT 3.6 Increasing Reliability and Maintainability

books) had better set the target high. Those writing for Reader Y (people who are used to complicated books and are not afraid of them) can set it somewhat lower.

Wherever the level is set, though, it is one of the best predictors of how often the book will fail. It is a concept complementary with *maintainability*: how long it will take to correct the problem and get the user going again. For, just as tangles and loops complicate the job of the user, similarly they complicate the job of finding out what is wrong with the manual and changing it.

The subtlest and most insidious problem with tangled, unusable manuals is that a **change on one page produces surprising, unanticipated errors elsewhere.** When, for example, a chart is removed from a manual, are we sure we have also adjusted every sentence that mentions the chart? If we have changed the name of a certain report, are we sure it has been changed everywhere it appears? If we have relaxed a restriction, has that relaxation been mentioned in every relevant procedure?

Getting changes into the manuals themselves can also create a mess. Some books (maintainable ones) accept supplements and modifications gracefully. Others fight them. In some DP centers, where the Standards Manual has been revised frequently, there are not two identical versions of the book to be found!

A Structured Approach to User Documentation

Chapter

4

"Cultures": How User Manuals Get Written

4.1   Two Ways to "Write" a Manual

There are two broadly different ways to write a manual. The first is to *compose* it, crafting the sentences and paragraphs while they are being written, as would a "writer" working on a script. The second is to *engineer* it, preparing a series of increasingly finer specifications until, at last, a manual "drops out."

When people think of "writers," the image that usually comes to mind is of a person slaving over the sentences in a manuscript; there are bursts of inspired scribbling, followed by intense editing and reworking, interrupted by long waits for the next inspiration.

There is a similar stereotype for computer programmers: people who think with their hands on the keyboard – trial and error, inspired guesses, flashes of genius.

These stereotyped programmers and writers do, in fact exist. Programming and manual-writing, you see, are two of the very few complicated projects that can be carried off in this loose, unplanned, "artistic" style. Indeed, computer programmers and technical writers are among the very few people I know who would, without hesitation, invest three or four person-months of effort on a nearly unspecified project and hope for it to turn out well.

Of course, there are very few professionals in any field who work entirely without planning. As stereotypes will do, this one exaggerates and simplifies. Still, most writers of manuals, including more than a few professional technical writers, work with even less specification than the people in any other technical profession. To write a long book from an ordinary outline – an outline that uses the same conventions the writer learned in grade school – shows little appreciation for the virtues of engineering.

As Exhibit 4.1 shows, the artist stereotype puts relatively little effort into planning. The main push is in the drafting stage; thereafter, the biggest effort is applied to "patching" the problems in the manual.

Exhibit 4.1 also shows the distribution of effort for the engineer stereotype. Here, most of the effort is in the planning – definition, design, modeling. The draft is merely the implementation of the design, not the creation of the product.

A complementary distinction between the two cultures occurs when other people get involved. The artists do not want to show the work until it is "ready." In contrast, the engineered manual is discussed extensively – and criticized and revised extensively – at several intermediate stages, before the author's ego is too deeply invested in the work.

The names "artist" and "engineer" are not supposed to suggest that all artists work without planning or that all engineers are so perfectly

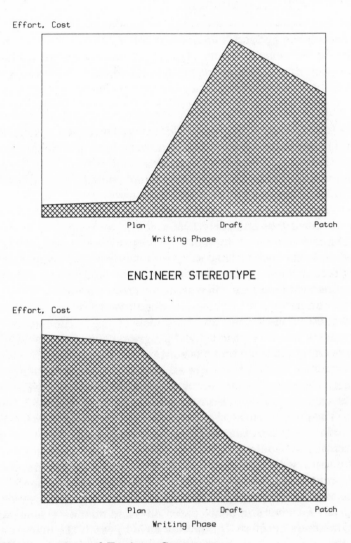

EXHIBIT 4.1 Artist and Engineer Stereotypes

disciplined. In truth, many professional writers prepare elaborate plans before they commit themselves to a draft and many engineers solve problems with casual trial and error—although computer programmers are likely to call it "prototyping." Rather, the purpose of the distinction is to emphasize that certain professions can be practiced with either "culture," and, moreover, that, when the projects get complicated and the stakes get high, the artist should yield to the engineer.

4.2   What Documentors Can Learn from the History of Programming

Programming has evolved from an informal craft into a formal branch of engineering. Fortunately, nearly all the tools that have been developed to improve the craft of programming can be adapted to writing user manuals. Of these, the most important is *top-down implementation and testing*.

In a 1979 issue of *Computerworld* Robert Perron wrote that "a comparison of programming in its early days to technical documentation in its present state yields some striking similarities." That is, people writing manuals in the 1980s often resemble people writing programs in the 1960s. They are prone to the same errors and "cultural" habits, and their products (programs or manuals, respectively) have similar flaws.

In the late 1950s and early 1960s, computer programming was an exotic, if not eccentric, profession. The people working in it were drawn by aptitude and passion. They were *not* trained by the schools and colleges and they were *not* treated like ordinary white-collar employees. Programmers worked alone, like artisans, often without much supervision, sometimes without budgets or deadlines to worry about.

Two factors, more than anything else, changed the nature of the programmer's job. First, the typical program became too large for one person working alone, ending the solitary luxury and independence of the programmer. Second, the major expense of programming shifted from inventing programs to maintaining them, and with that shift came the realization that most programs were disorderly, tangled, unmaintainable messes. (The term "spaghetti code" came into the programmer's slang.)

In today's organization, however, it is more likely to be the *writer* who is treated with deference, who works with little supervision and not much budgetary constraint. Today, for example, **most companies have no idea what it costs to write a page of user documentation.**

But, just as complexity, size, and maintenance problems made the old way of programming obsolete, so are they making the old way of writing manuals obsolete. If manuals have to reach the market at the same time as the technology, then the writing must be managed, budgeted, scheduled, and produced by teams of writers working in parallel. And if manuals have to keep pace with systems that are revised every few months, then the manuals have to be modifiable.

And what lessons have the programmers learned that the documentors should also learn? First, the single most important principle of software engineering: **the cost of detecting and correcting a problem rises exponentially as a function of how late in the development cycle the problem occurs.** That is, what costs a few minutes or a few dollars to fix at an initial planning session can cost hundreds during design, thousands during implementation, tens of thousands during distribution and operation.

To become an engineer, then, either a programmer or a documentor must adopt an attitude that may come hard at first: **an eagerness to find errors.** Usable and reliable technology is the result of testing, and the function of testing is to make things fail. Anyone who hopes that the test will show no flaws, that the specification will generate no arguments, that the outline will raise no questions – anyone who hopes that errors will come up later (rather than sooner) is asking for expensive problems and poor quality.

In sum, then, what documentors must learn from the history of programming is the craft of *top-down implementation and testing*:

1. The sooner an error or problem is detected, the cheaper and easier to correct it. Therefore, privacy and informality in the early stages of a manual are quite expensive.

2. The most serious problems in a complicated product are usually in the connections and interfaces, not in the units or modules. Therefore, the cost-effective way to develop a manual is to build it top-down, to assure the right mix of documents and the right content and sequence within each document – before the draft.

3. Unless a project has been designed top-down, it may take longer for several people to do the job than for one person working alone. Therefore, when manuals must be prepared on an accelerated schedule, they must be written to a detailed, top-down model.

4. It usually costs much more to maintain and support a complicated product or system than to design it simply in the first place. Therefore, the claim that there is not enough time and money to develop high-quality, maintainable manuals is nearly always false.

4.3   Goals for an Effective Documentation Process

What is needed, then, is a documentation process that will improve the suitability and appropriateness of the availabile documents; reduce the skips, jumps, and detours; enhance clarity, readability, and reliability; and make the publications easier to maintain.

Even those firms that have begun to conduct hard-headed laboratory tests of their user manuals (complete with human factors psychologists and formal test protocols) will soon learn that errors in a complete draft are far more recalcitrant than errors in an outline. Consider the analogy with those firms that do aggressive unit tests of their program modules but just cannot seem to integrate the tested modules later on. What these firms seem to overlook is that program modules, like book modules, must be integrated *before* they are written, not after. And that, as mentioned earlier, the most agonizing problems in writing or reading manuals are in the links and connections, the interfaces, not the individual units or pages.

So, a documentation process that benefits from the last decade of software engineering will achieve the five goals listed below.

It will improve the "fit" between user documents and the needs and convenience of the users. The method, then, must include a way of aligning the material to be written, the publications and other information products, with the users of those products. In other words, the process must be driven by the particular characteristics of the users and operators and their peculiar interests in the system, rather than a one-size-fits-all standard for user publications. More simply, the process must recommend ways to define a logical mix of books and materials, a Document Set. Furthermore, the proposed contents of this set must be *testable as a proposal.* That is, it must be possible to review the plan and find strategic errors in the making, well before the plan is turned into manuals and disks.

It must reduce the skips, jumps, and detours in the publications or other products. Even though modern word processing makes it easier to move around blocks of text, it is still inescapably true that, once a book or manual is completely drafted, the manuscript develops an inertial resistance to structural change. So, any effective technique for documentation will, necessarily, expose structural errors *before* the inertia of the draft takes hold. An effective documentation process will generate a series of increasingly more detailed models of the product. And these models – which appear between the outline and the draft, the interval when the artistic writer usually works alone – will be testable against clear measures of usability.

It will allow writers to work in teams and in parallel. The most common excuse for inadequate documentation is the claim that preparing it would delay the delivery or implementation of the system by several weeks or months. But this claim, far from being an adequate explanation for poor user documentation, is, rather, proof of the need for techniques that will allow user information products to be written by teams of people, working on well-defined chunks of the publications, *in parallel*.

Note that, without the right method, having writers work in teams can actually slow the process; with the wrong approach, two writers will take two or three times longer to write a book than one writer! So, an effective documentation process will organize the work into a set of manageable parcels, capable of *independent execution*, with costs and schedules that can be predicted and controlled.

More specifically, it will "decompose" the large job of writing into a set of small jobs, tiny documents, of easily estimated size and cost. Furthermore, all the links and interfaces between the tiny documents have been defined, explicated, and tested in the model. Thus, each of the small pieces can be written independently, without consulting the authors of the other pieces – so long as each author has access to the model of the whole publication.

It will enhance the clarity, readability, and reliability of the manuals and other information products. Of course, the flaws in the draft are important. The goal, though, is to solve every strategic problem and correct every structural flaw *before* the draft is composed. In that way, what will remain to be corrected in the draft are precisely those problems that lend themselves to editorial improvement: incorrect claims about the system, minor technical changes, unclear sentences, ambiguous paragraphs, cluttered or confusing illustrations, and so forth.

Furthermore, an effective documentation process will include formal standards for editing and will not rely on the artistic, intuitive, "stylistic" preferences of one person. For example, the prompt "Press F(4) to continue." is a backwards, unreliable sentence. A sound documentation process will flag and correct this sentence, whether or not the editor finds it personally unobjectionable, and whether or not any reader has trouble with it in a test.

It will generate publications and products that are maintainable and modifiable. Well-made documents will not have to be revised and supplemented as often as ill-made documents. But when they do need revision, the process will be more straightforward and less subject to confusion and misinformation.

5

A Structured Process: Five Steps in Developing Manuals

5.1   What "Structured" Means

5.2   What "Modular" Means

5.3   An Overview of the Process
    5.3.1   The Flow of Data Through the
           Documentation Process
    5.3.2   A Work Breakdown of the Process for
           Developing Manuals

5.1   What "Structured" Means

The term *structured* can be applied to user documentation in two main ways. First, the process for developing user manuals is characterized as a "structured process." Second, the publications themselves are often called "structured documents." Unfortunately, the word structured is used so often and so casually these days that it is necessary to stop and define what it means.

By *structured*, I am referring not to its overworked conversational meaning, in which it is a loose synonym for disciplined or organized. Rather, I intend it in the same sense as used by computer scientists and software engineers in such expressions as "structured analysis," "structured design," and "structured programming."

In all three uses, structured refers to a certain process or method, well put in the definition of *structured analysis* below:

> A formal, top-down decomposition of a problem or process into a model that offers a complete, precise description of what the problem is . . . used as a basis for the coding . . .
>
> — Sippl and Sippl, *Computer Dictionary & Handbook*
> (Indianapolis: Sams & Co.), 1980, p. 529

First, structured analysis is *formal*, which means that it is public, explicit, and rule-abiding; that is, a process cannot be considered structured if it is intuitive, private, and conducted without rules or guides. Next, it is *top-down*, which means that it starts with the biggest picture possible, the whole system with all its interfaces, and adds overlays of detail in successive stages. At each further level of detail it is *tested*.

Many people in the DP industry mistakenly believe that top-down analysis consists in breaking big ideas into increasingly smaller ideas. That is what is meant by the next key term, *decomposition*. Although structured analysis requires decomposition (disaggregating big things into smaller things), it first requires a representation of the entire system. In structured technology, we know that the parts fit into the whole before we define the insides of the parts.

The next key word in the definition is *model*. Put simply, in structured methods we build models of a product before we build the product itself. Why? Because it is vastly cheaper to build and change models than it is to change the finished product.

After structured analysis comes *structured design*:

> The art of designing the components of a system and the interrelationship between those components in the best possible way. Or, the process of

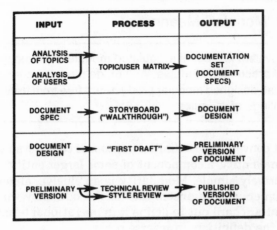

INPUT	PROCESS	OUTPUT
ANALYSIS OF TOPICS ANALYSIS OF USERS	TOPIC/USER MATRIX	DOCUMENTATION SET (DOCUMENT SPECS)
DOCUMENT SPEC	STORYBOARD ("WALKTHROUGH")	DOCUMENT DESIGN
DOCUMENT DESIGN	"FIRST DRAFT"	PRELIMINARY VERSION OF DOCUMENT
PRELIMINARY VERSION	TECHNICAL REVIEW STYLE REVIEW	PUBLISHED VERSION OF DOCUMENT

EXHIBIT 5.1   Applying Structured Methods to User Documentation

deciding which components interconnected in which way will solve some well-defined problem.

> — Yourdon and Constantine, *Structured Design*
> (Englewood Cliffs, NJ: Prentice-Hall), 1979, p. 8.

Notice that a product designed this way has only two things in it: components and the relationships between them; modules and interfaces; nodes and edges; units and links. And, because there are only these two kinds of entities, it is usually possible to describe a structured product or system with only a simple diagram containing blocks or circles for the modules (nodes, components, units) and arrows or lines for the connections (linkages, interfaces, edges).

Again, the reason for making such diagrams — especially in the planning of a book — is to **find flaws and problems while it is still cheap and easy to correct them.**

Ultimately, of course, the real benefits to the developer are at the maintenance phase, where all changes will consist simply in replacing or adding one small module or unit, and where the effects of making that change will be predictable from a study of the design.

The point to be stressed here is that the same structured methods used to make programs and systems more cost-effective and maintainable can be applied directly to the job of devising and writing user manuals. And with similar benefits. Further, if the *process* is structured, then the *products* — the publications — will also be structured. They will consist of many small components, modules, connected in a way that makes the book as usable as possible.

5.2   What "Modular" Means

A structured system is a set of articulated modules; a structured publication is a set of articulated modules. Well-made modules are cohesive and predictable; well-designed modular products are free from excessively complicated couplings of modules.

The most conventional definition of *module* calls it a small, independent functional entity, a component of some larger entity. This definition is deceptively simple. Modularity is an elusive, almost mystical quality. Designers and engineers can "feel" modularity when they get close to it, but are hard put to define it in operational terms. Consider the parts of the definition, in reverse order.

Modules are functional. Modules are not just parts of something larger; they are *functional* parts; they do something that can be described. A module performs some task, converts data from one form to another, more-usable form. Moreover, well-made modules usually perform a *whole* task, for example, sorting all the accounts in a file into those that are current, past due, and extremely past due. A well-made module is also predictable: the same inputs arriving under the same conditions will generate the same output; there is no "internal memory" in the module that would change the input/output patterns.

Modules are independent. Modules are not dependent on their context; a module with a particular function will perform that function in more than one setting. In some views, each module becomes part of a large set or library of reusable modules; eventually, designers can create systems or products merely by shopping from the list of available modules and organizing them in a novel way.

Modules are small. The least precise part of the definition refers to the size of modules. To say that modules perform only one function fails to limit them precisely. People who work with modular software eventually decide that long theoretical arguments about whether something is one module or more than one are a waste of time. As a practical measure, then, most people who work with structured technologies limit **the maximum size of a module.** In data processing, the limit is usually a certain number of code statements; in publications, it is a certain number of pages.

Indeed, one of the interesting parts of developing modular systems or publications is playing with the size of the modules. As modules get larger, they get less cohesive (more than one function); as they get smaller, though, the couplings and connections become more complicated. In modular publications, these couplings manifest themselves as references to other pages in the book. And, the central argument of the book you are reading is that these sorts of design decisions – such as trading-off module cohesiveness for inter-module complexity – can be applied directly to the development of more usable user documents.

The first mature attempt to treat documentation this way can be found in the invention of a group of publication engineers working, at that time, for the Hughes Aircraft Company. Their process, a form of "storyboarding" adapted from the motion picture industry, and their "modules," a series of two-page spreads, are described in their seminal work on the subject:

> Tracey, J. R., Rugh, D. E., and Starkey, W. S. *STOP: Sequential Thematic Organization of Publications*. Hughes Aircraft Corporation: Ground Systems Group, Fullerton, CA, January 1965

The earliest journal publication is:

> Tracey, J. R., "The Effect of Thematic Quantization on Expository Coherence." *IEEE International Convention Record*, Paper 9.4, 1966

Modular manuals benefit not only the readers of manuals, but also the designers and writers. Working from modular, "structured" outlines, designers are able to predict the size and cost of publications, at the same time they are testing them for readability and accuracy. Furthermore, by breaking the long, complicated process of writing a manual into a set of small and independent tasks, firms can apportion the writing assignments to a great many people who can work in parallel, independent of one another.

Modular manuals are also a boon to "authors" – all those people we usually call on for raw "input" to the manuals. In the modular manual, these people can be transformed into "first drafters," each knowing exactly how much to write and exactly what points to cover.

Even writers working alone as "artists" benefit from modularization. They can work in short bursts, knowing that the little pieces will ultimately fit together well. But it especially benefits those who manage writing and supervise publication. Planning, writing, editing, and producing by module radically increase the control of the person in charge.

And, perhaps most important, effectively designed modular manuals are the most readable and "user-friendly" technical publications imaginable. Indeed, the reactions of readers to modular publications have done more to sell the concept than all the arguments in its favor by authors and developers. Although I have met a few technical writers who dislike modular manuals, I have never yet met a reader who does.

5.3 An Overview of the Process

The five phases of user documentation are: *analysis* (defining what information products are needed); *design* (building models of the products and testing them); *assembly* (forming a complete draft); *editing* (eliminating "language bugs" and technical errors); and *maintenance* (changing what needs to be changed).

Just as there is a necessary cycle or process for system development, so is there an appropriate cycle or process for user documentation:

1. *Analysis*, defining just what manuals and other information products the users and operators need. Obviously, the earlier in the life of the system the analysis takes place, the better. Ideally, the documentation analysis – which, in large projects, often culminates in a Publications Plan – should occur as part of the original system development plan. But it is never too late to analyze the remaining documentation need. And documents written without an analysis of what documents are needed usually fail.

2. *Design*, preparing detailed outlines of each manual or other information product. This phase starts with the preparing of conventional outlines, but then proceeds to the creation of "structured outlines" and "storyboards" – working models of the documents that can be tested and revised before the first draft is written. Recall: The most difficult structural and organizational problems must be corrected before the first draft is written.

3. *Assembly*, the conversion of the storyboard to a workplan, and the writing of the first draft. In the structured approach to documentation, writing the first draft is a little like writing the code in a structured program: that is, the authors do little more than fill in missing details, according to a strictly followed plan, the "storyboard."

4. *Editing*, testing the first draft for clarity, correctness, and "readability": the ease with which the text can be read. Note that, in this approach, questions of language and style are far more than matters of esthetics; rather, the purpose of this phase is to apply principles of editing that make the manual easier to use, and, therefore, less likely to cause a "failure," defined as what occurs when an operator or reader is unable to work the system because of a bug in the manual. In many cases, this phase culminates in a test with "live" readers.

5. *Maintenance*, keeping track of what needs to be changed in the information products and, when appropriate, making the changes. Because all manuals are flawed or out-of-date (without exception), the

last phase of documentation is to monitor what should be added, removed, replaced, or repaired. Actually, the subtle craft of maintaining user documents is, first, to know what changes must be made, and, second, to be able to distribute and incorporate those changes in a manner that does not generate confusion and additional errors.

All five phases are necessary for effective, usable documentation. Organizations that skip the first two (or three or four) phases will produce documents with all the expensive functional and structural flaws of undesigned computer programs and sloppily engineered machines.

User documentation is rarely effective when it is approached as "cleaning up" old problems. People who start at the end – editing and maintaining old publications that never worked well in the first place – are usually wasting their efforts. Just as it makes no sense to automate an inefficient manual procedure, it makes little sense to invest time and money in the endless repair of poor publications.

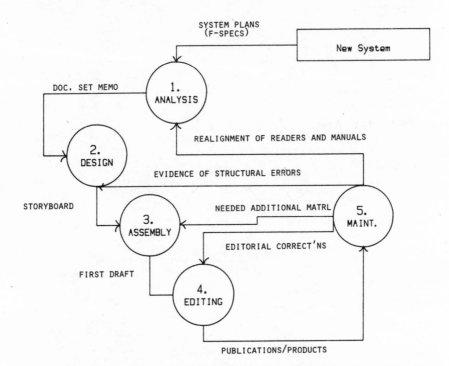

EXHIBIT 5.3   Data Flow Diagram (DFD) for Developing User Documentation

5.3   An Overview of the Process
5.3.1   The Flow of Data Through the Documentation Process

The five small diagrams in this section are the "children" of the Data Flow Diagram on the previous page. Each shows, at one lower level of detail, the activities in each of the five phases. (Data Flow Diagrams show data moving through arrows and being transformed in the nodes or "bubbles.")

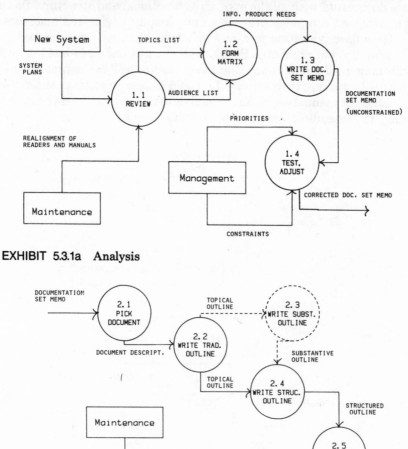

EXHIBIT 5.3.1a   Analysis

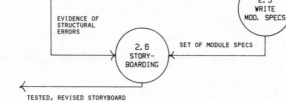

EXHIBIT 5.3.1b   Design

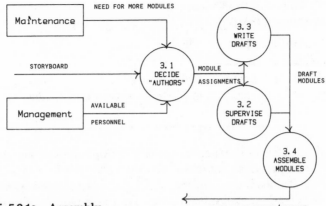

EXHIBIT 5.3.1c   Assembly

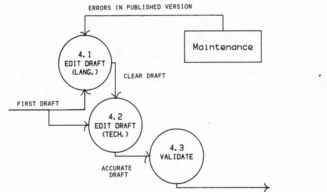

EXHIBIT 5.3.1d   Editing

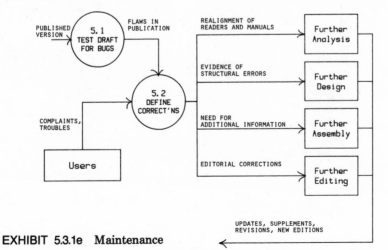

EXHIBIT 5.3.1e   Maintenance

5.3   An Overview of the Process
5.3.2   A Work Breakdown of the Process for Developing Manuals

---

The exhibit in this section shows a simple hierarchical breakdown of the tasks needed to develop usable user manuals and related information products.

Exhibit 5.3.2 lists the tasks necessary for high-quality user documentation. Although it does not include every task (such as cost estimating or printing), it does include every activity needed to realize the "structured" method advocated in this book. That is, anyone wishing to skip some of these tasks or perform them in another sequence must have a persuasive justification for doing so.

Note also that the exhibit does not address the question of who should perform each task. Because there is so much variety in the staffing and organization of firms and agencies who write user documentation, it is impossible to say who should do what. Rather, the next six chapters discuss each of the tasks in some detail and advise on the kinds of skills needed for each.

EXHIBIT 5.3.2   Work Breakdown for Developing User Documentation

(1.0)   ANALYSIS: Developing the *Documentation Set Memo*

(1.1)   Review the system/product description
(1.2)   Review the intended audiences and users
(1.3)   Prepare a list of features, tasks, and topics
(1.4)   Prepare a *topic/audience matrix*
(1.5)   Define a set of manuals and other information products
(1.6)   Assemble the unconstrained *documentation set memo*
(1.7)   Test and adjust the plan, as needed

(2.0)   DESIGN: Developing a Detailed Model

(2.1)   Pick a document and write a traditional (topical) outline
(2.2)   Write a substantive outline (optional intermediate step)
(2.3)   Write a *structured outline*
(2.4)   Test and adjust the outline, as needed
(2.5)   Write a *module spec* for each item in the structured outline
(2.6)   Test and adjust module spec, as needed
(2.7)   Assemble the storyboard model of the product
(2.8)   Review, test, and revise the model until acceptable

(3.0)   ASSEMBLY: Generating the Draft

(3.1)   Decide who will write each module
(3.2)   Assign writing tasks
(3.3)   Coordinate, collect drafts

(4.0)   EDITING: Testing and Refining the Draft

(4.1)   Edit drafts for clarity and readability
(4.2)   Edit drafts for technical accuracy
(4.3)   "Validate" the draft for usability (test with real users)

(5.0)   MAINTENANCE: Supporting and Updating the Manuals

(5.1)   Test the manuals and other information products for bugs
(5.2)   Review product changes and enhancements
(5.3)   Prepare updates, supplements, or new editions
(5.4)   Distribute maintenance information products

Chapter

6

Analysis: Defining What Publications Are Needed

6.1   Documents Always Come in Sets

There are certain fundamental questions about the content of a publication that cannot be answered without defining the larger set of documents and information products of which it is a part.

The proper way to begin user documentation is to define a *set* of information products (books, reference cards, videotapes . . .) and then to define a specific function and scope for each item in the set. Why? Because to define what a thing is, you must also define what it isn't. The surest way to clarify the purpose of a publication is to contrast it with other, adjacent publications.

The systems approach to a problem consists in viewing it as part of a larger problem. Before we can know what to put into Book A, we must know why there are any books at all, what they do as a group, and what they do as individuals.

Indeed, the most appropriate way to define a set of user documents is to think of them as part of a larger set of items called *information products*, including not only publications but also audiovisual products, online tutorials, and the whole range of teaching and reference media.

Furthermore, the most appropriate way to define the needed set of information products is to view it as part of a still-larger entity called *user support*, which contains not only information products but a full range of user services (see Exhibit 6.1).

Notice also that there are even tradeoffs between the quantity of information products needed and the quantity of services needed; high-quality information products can reduce the need for training, consulting, and maintenance. In fact, that is a main cost justification for investing in user documentation.

Put simply, the time to decide the scope of a particular publication is not during the writing of the outline and certainly not during the writing of the draft. The time for definition is *before* the outlines are written. The time to argue about whose information needs will be served is at the beginning; the time to argue about whether two publications will overlap is before either of them has been outlined; and so forth.

Yet, as obvious as this principle may seem, most writers of manuals ignore the issue. Like programmers eager to produce some code, the documentors are eager to produce some text. Rather than debating the scope of a document before it has been written, they prefer to write a draft and worry later about whether it will or won't serve certain readers.

And the consequences are the same. The finished draft, like the coded program, develops inertia; its author becomes its defender. What users or customers need has far less influence than what has already been written and paid for.

Right now, there are hundreds of skillful writers struggling with undefined and misconceived publications. Unfortunately, these writers think that their problems are *within* the publication. Actually, the problem is strategic: the lack of an information products plan.

It avails us little to be competent writers if we write the wrong manual. And the only way to be sure of what a manual is, the only way to know what to include and "include out" is to differentiate each product from the others in the set.

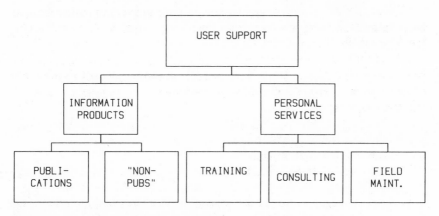

EXHIBIT 6.1   User Manuals in Context

6.2   Forming the Document Planning Team

Analysis begins with the appointing of the Documentation Team. Its mission is, simply, to discuss the documentation needs of System S and prepare the Documentation Set Memo: a list of recommended documents and a thumbnail description of each. To be effective, the team should contain, first, an expert on the technology, second, an expert on the uses of the technology, and, finally, a coordinator to bring these two perspectives together.

Most firms get nowhere with their documentation problems until they empanel a team whose mission is to name and describe the documents that shall be developed.

Deciding what documentation is needed is too important a decision to be made casually or by default. Also, it is too important to be made by one person. Rather, defining what documents are needed calls for at least three perspectives:

- the technology expert
- the application expert
- the document coordinator (referee)

The *technology expert* is the person who knows the most about the design and inner workings of the system. Known variously as a systems analyst, lead designer/developer, project manager, or just engineer, the technology expert must speak for the system. If the system has already been developed, he or she must be most knowledgeable about its features and characteristics. If it is about to be developed, he or she must be in charge of the functional specification or general design.

The *application expert* (often referred to simply as the "user") must know what the system is for. Users, end users, operators, manufacturers, technicians, auditors, marketing managers, trainers, customer relations people – any of these may be the most likely participant on the Documentation Team. The task of the application expert is to remind the team members – as often as necessary – that the system will have to be used and operated. And that the users and operators (and supervisors, administrators, or even salespeople) often have no interest in the inner workings of the system, frequently have little or no technical education, and are almost always interested exclusively in what the system can do for them.

In most cases, these two points of view will conflict – just as user requests and analyst responses almost always conflict. Thus, the third member of the team, the *document coordinator* has the task of managing

EXHIBIT 6.2   Document Planning Team

Perspective	Candidates	
• Technology	• Systems Analyst • Lead Designer/Developer • Project or Product Manager • Hardware Expert • Software Expert	
• Application	• User • Operator • Manufacturer • Technician	• Auditor • Marketer • Trainer • Customer Relations
• Coordination	• Documentor/Editor • Business Systems Analyst • Liaison/Coordinator • QA Reviewer • Technical Writer • Publications Engineer/Manager	

the conflict and deriving a consensus from the others. The coordinator— known variously as documentor (even "documentalist"), "business systems analyst," liaison, Quality Assurance rep, technical writer, or even publications engineer—must produce the actual memo or list. He or she must listen carefully to the two perspectives, follow some procedures that will be discussed in a while, and produce the Documentation Set Memo for System S.

Note that, in this plan, the person responsible for documentation is a manager/coordinator, involved in the earliest planning, rather than a copy editor brought in to clean up untidy drafts. Note also that the job of defining documentation needs cannot be left either to the technical expert or to the application expert alone; in general, neither sufficiently appreciates the other's point of view. And in practice, they often find communication very difficult.

Ideally, there should be three members on the team, one from each category. There may be more if the system is extremely complicated or has an unusual mix of users. Be careful, however, that, if there is more than one technical expert, the appropriate power prevails. If there are hardware and software experts, for example, there should be someone who speaks for the *system*. Failing that, the coordinator must be vigilant in preventing any of the limited technical specialties from overburdening the documentation plan with irrelevant technical detail.

Also beware of the two-person documentation team, which, for the sake of "efficiency," suppresses the needed conflict. And be especially cautious if the definition of documentation needs is the work of only one person.

6.3   Listing the Features and Topics

The technology expert on the team analyzes the components, features—the *topics*—that need to be written about. Although there are countless schemes for categorizing the aspects of a system, and although a task-oriented method is best, the particular approach is less important than the completeness and fineness of the analysis.

In analyzing the documentation needs for a particular product, a critical task is to decide just what the product is. Just what is worth knowing about it.

The topic analysis (or sometimes functional analysis) is the job of the system expert on the documentation team. Although the breakdown will inevitably be influenced by the other members of the team, it is still the system expert's job to describe the structure and morphology of the system itself.

Systems can be described variously by talking about their physical components, their design, their technology, their operations, their applications, or their benefits.

Exhibit 6.3a shows part of an illustrative breakdown of a system by its components, the things it contains. Frequently, such a breakdown has already been prepared as part of the system planning.

In the 1980s, there is much talk in documentation circles of *task-oriented manuals*, which are organized according to the tasks performed by the intended reader. (See Exhibit 6.3.b.) Task-orientation—which is usually contrasted with product-orientation—is an application of what the instructional technologist calls "skills analysis" and what some social scientists call "activity accounting." Put a bit too simply, a task-oriented manual supports users in exactly what they do and refrains from general discourse on the characteristics of the system. Moreover, in the IBM variant of this concept—the source of several of the most interesting papers at recent meetings of the International Technical Communication Conference—there is an attempt to develop a "universal task architecture," a standard classification scheme for all possible tasks and all possible software products or systems.

As will become clear in a while, the best way to eliminate loops and detours from a manual—to raise its usability—is to make it task-oriented for a well-defined audience. Remember, though, that even if this initial breakdown of topics is *not* task-oriented, there are still opportunities later in the documentation process to incorporate this idea.

Still another approach, even harder for the typical analyst or engineer, is to break the system down according to its benefits or selling themes, as shown in Exhibit 6.3c.

EXHIBIT 6.3a
Topics Listed by
Components

INPUT DEVICES
OUTPUT DEVICES: VDTs
OUTPUT DEVICES: PRINTERS
OUTPUT DEVICES: PLOTTERS
DISK STORAGE
TAPE STORAGE
RANDOM ACCESS MEMORY
READ ONLY MEMORY
MICROPROCESSOR CHIPS
OPERATING SYSTEM
UTILITIES

EXHIBIT 6.3b
Topics Listed by Task

PRODUCT SELECTION/EVALUATION
PLANNING/SITE PREPARATION
INSTALLATION
RESOURCE CONFIGURATION
CUSTOMIZATION/DEFAULTING
WRITING APPLICATIONS
TROUBLESHOOTING/MAINTENANCE
FILE CONVERSION
CREATING SITE-SPECIFIC TEMPLATES
COMMUNICATING WITH OTHER SYSTEMS

EXHIBIT 6.3c
Topics Listed by
Selling Themes/Benefits

KEYBOARD REDEFINITION
USER-DEFINED FUNCTION KEYS
TYPING PRODUCTIVITY TOOLS
WORD PROCESSING "MACROS"

There are several orthogonal paths of attack in defining features and topics. Although task breakdowns are nearly always more useful for documentation than physical breakdowns, any scheme will do as long as the analysis is fine enough.

How fine? The topics must be small enough and clear enough so that team members can ask the following general question: Is Topic T necessary or important for Reader R? Yes or No? If the topics are defined too broadly or vaguely, then the analysis must be refined.

6.4   Analyzing Audiences: Users and Readers

One of the key tasks in defining specialized documentation needs is to analyze the audiences: groups of users and readers that need to know about the system. This analysis or breakdown is the responsibility of that person or persons on the documentation team who either represent the user/operator unit or are supposed to be expert on the uses of the system.

Let's define an *audience* as the intersection of interest and background. If two readers have the same interest in a system and the same technical or business background, they are part of a single audience. People with different interests in the system—that is, different roles, different applications—are in different audiences. And, similarly, people with different technical backgrounds, even though they use the system for the same purpose, should be regarded as different audiences.

The variety of users corresponds to the expanding variety of the products and systems. And, increasingly, the newer audiences tend to be people with less technical background and with slight experience in computer technology.

The breakdown of users and readers need not be excessively fine. It usually takes only between three and ten categories to define the relevant groups of users. (The breakdown in Exhibit 6.4 shows ten categories and is purely illustrative.) There are exceptions, of course, systems

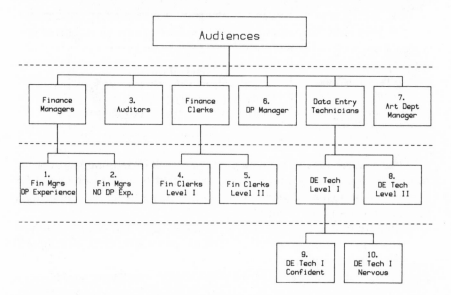

EXHIBIT 6.4   Analyzing Audiences

with more diverse audiences, more than ten categories. It is better to err on the side of having too many audiences, because later steps in the analysis will eliminate the redundant and unnecessary categories.

The simplest breakdown is by occupational specialty or job title. (Strictly speaking, people with the same background and interests are in the same audience, even if they happen to have very different job titles.) A second cut at the list of audiences is to differentiate within each of these categories, to distinguish those with long experience from those with little. A recurring problem in user documentation is the practice of writing one manual to serve both the experienced veteran and the brand new trainee—just because they happen to have the same assignment. At this stage, then, it is wise to break the larger occupational categories into subgroups with different levels of experience or types of background.

In some cases, especially when much of the documentation is for teaching users and operators how to work a system, it may also be necessary to include a third dimension for defining audiences: *nervousness*.

Usually this nervousness dimension is so closely correlated with inexperience and lack of technical education that there is no need to account for it in the breakdown of audiences. If, for example, you have an audience called "data-entry-clerks with high school education," you can expect most of them to be nervous and insecure. And if you have a group of mechanical engineers using a system for Computer Assisted Design, you can expect most of them to be industrious and not easily scared by the system or the books.

The breakdown of readers is critical. **It is a serious strategic error to write your documentation as though it were one compendium of material aimed at a universal audience.** The task at this stage is to make sure that every important user, operator, manager, programmer, or other person who comes in contact with the system gets the documentation that he or she needs. For now, we want the "worst case": the finest possible breakdown of users and readers. To decide what documents are needed, we must think hard about who needs them and why.

6.5   Forming the Topic/Audience Matrix

With the two breakdowns completed, the document coordinator on the team prepares the topic/audience matrix. The team as a whole then analyzes the intersections of topic and audience, indicating the points at which particular topics affect particular audiences.

The team transfers the two breakdowns to the reader/topic matrix and decides which topics are of interest to which users.

The decision is usually an easy consensus. When there is disagreement, the easiest solution is to include the disputed topic. (In general, the safest way to resolve choices about documentation needs is to provide more, rather than less, information.)

In Exhibit 6.5, the topics are refined enough to allow simple yes/no choices. That is, each topic is small enough so that a particular audience needs to know all of it or none of it.

Interestingly, a matrix very much like this one is often prepared in corporate training departments. Sometimes it is called a "skills matrix" or "task matrix," used to define the training needs of various audiences. If your firm already has such an analysis in its training department, it may be useful in analyzing documentation needs as well.

What does the matrix mean? Is it just another of those time-consuming planning tasks that cause programmers to grow impatient and long to get back to their coding? Hardly.

The emerging pattern of checkmarks suggests the shape of several documentation products. Without it, there is a high probability that documentors may write the wrong books.

The matrix is a thinking tool. If there is a topic without a checkmark, it forces the panel to consider the possibility of an overlooked audience. If there is an audience without a checkmark, it means that the information needs of certain readers have been ignored.

Furthermore, the matrix often shows that people with presumably different interests have remarkably similar information needs. (Data Center Managers and Data Entry Clerks often receive similar checkmarks, for example.) Or, more important, it demonstrates that certain audiences have been neglected, or swamped with irrelevant information, or lumped together with readers whose needs are quite different.

Think of the long list of topics as the inventory of documentation materials; think of the list of reader groups as the consumers of those materials. The goal of this analysis, then, is to decide how the materials should be "partitioned" for the convenience and necessity of the consumers.

Topic/Audience Matrix for PRODUCTION SCHEDULER

	Corp Mgmt	Finan Mgmt	Manufac Mgmt	Book-keeping	Data Entry I	Data Entry II	Art Dept	Legal Dept	Secur-ity	Quality Assur
1 Proj Objectives	✓	✓	✓						✓	✓
2 Implement. Sched.		✓	✓						✓	✓
3 Cost Savings	✓	✓	✓						✓	✓
4 Improvd Quality	✓	✓	✓							✓
5 Site Prep			✓						✓	✓
6 Security Setup			✓					✓	✓	✓
7 Installation Proc			✓			✓			✓	✓
8 Customizing			✓			✓				✓
9 Initializing Files			✓			✓				✓
10 Loading Prod. Plan			✓		✓	✓				✓
11 Loading Daily Runs			✓		✓	✓				✓
12 Revsng Prod. Plan			✓		✓	✓				✓
13 Loading Error Data			✓		✓	✓				✓
14 Status Reports	✓	✓	✓	✓	✓	✓				✓
15 Prodctn Forecast	✓	✓	✓	✓		✓				✓
16 Unit Cost Analysis	✓	✓	✓	✓		✓				✓
17 What-If Alterntvs	✓	✓	✓							
18 Progress Graphics			✓	✓		✓	✓			
19 Forecast Graphics			✓	✓		✓	✓			
20 Cost Graphics			✓	✓		✓	✓			
21 Accounting Intrfc		✓	✓	✓				✓		✓
22 Contracts Intrfc		✓	✓	✓				✓		✓
23 Data Entry Errors			✓		✓	✓			✓	✓
24 Reporting Errors			✓		✓	✓			✓	✓
25 Failure Procedure			✓			✓			✓	✓
26 Program Directory			✓			✓			✓	
27 Data Dictionary			✓			✓		✓	✓	✓
28										
29										
30										
31										

EXHIBIT 6.5   Topic/Audience Matrix

6.6 Partitioning: Specifying Boundaries and Overlaps

The team studies the matrix of topics and audiences to see if it can find patterns or clusters that mark off individual documents. At one extreme, the planners may decide to prepare a single, encyclopedic reference manual—with the matrix as an index for the reader. At the other extreme, they may decide to have separate volumes for each audience or each topic. Somewhere between these extremes lies the best solution; the ideal partitioning is determined by tradeoffs such as those shown in the exhibit.

As in defining the boundaries of systems, there are many decisions and tradeoffs—and often arbitrary rulings—involved in defining and delimiting the documentation products.

Various factors are played off against one another as the members of the team devise the most cost-effective mix of books, audiovisual materials, and other information products. Of course, few organizations ever produce more products than they planned, tending to partition toward the encyclopedic end of the continuum.

As Exhibit 6.6 shows, there is something to be said for and against each strategy. Individualized documentation, up to the point of having alternative versions of the manuals for ten or more audiences, has many communication advantages. It results in publications that are tailored

	INDIVIDUAL EXTREME	ENCYCLOPEDIC EXTREME
ADVANTAGE	• TAILORED TO THE USER • ACCESSIBLE, SHORT • CONFERS RECOGNITION • PROTECTS SECURITY • PREVENTS ABUSE • WORKS AS A TRAINING MANUAL • SAVES MONEY (OCCASIONALLY)	• EFFICIENT, NOT REDUNDANT • SIMPLIFIES PLANNING • SIMPLIFIES PRODUCTION • SIMPLIFIES MAINTENANCE • PERMITS LEARNING • ELIMINATES RIVALRY • YIELDS A GOOD ONLINE SYSTEM
DISADVANTAGE	• COMPLICATES MAINTENANCE • UNDERESTIMATES USER • PREVENTS LEARNING • COSTS MORE (USUALLY)	• LONG, HEAVY BOOKS • INTIMIDATES USER • BURDENS TRAINERS, SERVICE PERSONNEL • DEMANDS SOPHISTICATED INDEX

EXHIBIT 6.6   Tradeoffs in Deciding How to Partition the Documents

precisely to the interests of the readers, thereby freeing them from searches and detours to other publications. It generates shorter publications, which can even have a prestige associated with their ownership – something not possible when everyone has the same version.

Short, individualized manuals also protect the security and confidentiality of material by restricting the access of certain readers. Similarly, they help us to prevent certain operators and users from trying procedures or features they have not been cleared (or taught) to use. Occasionally, it is even cheaper to have several versions. Sometimes we need hundreds or thousands of copies of a short publication but only a few copies of the longer one.

Usually, though, individualized documentation is more expensive, and it can be extremely difficult to maintain. Obviously, it's hard enough to keep one manual current, let alone several alternative versions of it. Individualized versions can also sometimes underestimate the abilities of the audience members, prevent them from learning skills that would increase their value, or even force them to consult several documents.

Most documentors, of course, do not analyze their documentation needs in this way. They think of user documentation as *one entity*, one file of literature and data. Sometimes this encyclopedic manual is the right choice; for simple systems with a homogeneous set of users, a single manual may be best. But, usually, the single manual is an expedient choice, a way of simplifying the documentors' planning and production and reducing short-term cost, without much regard for its usefulness to the readers. Furthermore, many of the firms that purvey encyclopedic documents neglect to include an index. An encyclopedia without an index and a system of cross-references is nearly impossible for readers to use. Yet, it still does not automatically follow that the more manuals the better.

One of the most interesting paradoxes in planning user documentation is that the encyclopedic manual – the most cumbersome and inaccessible to the typical user – is perfect for an online system of documentation. The difficult search routines, the cross-references and detours, the trimming away of what is uninteresting or irrevelant – these and other processing burdens are borne by the system instead of the user. The search programs are "transparent" to the reader, which, in computer parlance, means "invisible."

6.7   Excerpt from a Documentation Set Memo

After the deliberations of the Documentation Team, the *document coordinator* prepares a memo describing the plan agreed to by the committee. The plan contains a profile of each document and other information product that will be developed.

The Documentation Team prepares and distributes the Documentation Set Memo, which contains a plan for all the documents that will be written during the rest of the life cycle of the system. Alternatively, they may choose from a longer standard documentation set those products to be written for the current project.

For each document or information product, the team defines:

- *A Title* – a descriptive name for the publication, which represents its scope and audience.
- *A Control Number* – a code ID that allows all concerned to monitor the progress and costs of the document.
- *The Medium* – whether the document is to be a book and, if so, the form and materials (e.g., loose-leaf binder, signature staple, comb-bound); if not, the non-book medium to be used (e.g., video, poster, placemat).
- *The Audience* – a brief description of the occupational categories for the intended readers, or, when appropriate, the names of specific organizations, units, or persons.
- *A Priority Rating* – a measure of the relative importance of each document among all the others in the same set. This is especially important when the planners suspect that there will not be enough resources for every item in the set.
- *Completion Date* – an estimate of when the product will be ready, or, for larger projects, estimates of several milestones, such as storyboard, first draft, first test with "live users," final art, and typography.
- *Responsible Documentor* – the person who must get the product done, on time, to specifications, and under budget; manuals and publications hardly ever get written until some person feels responsible. Note that the title "writer" does not convey the job of this person; in fact, the responsible documentor need not be one of the writers.
- *Budget* – a description, in as much detail as is appropriate, of the personnel and non-personnel resources assigned to the project. Although user documentation is expensive, there is no gain in trying to "fold in" the costs with the other aspects of system development.

Title: Comprehensive User Guide for PRODUCTION SCHEDULER

Medium: Loose-leaf   Control #: PS-01   Priority: A

Audience: Manufacturing Management; Quality Assurance; Security Dept.

Responsible Documentor: Hemingway   Completion Date: Mar 85

Sections/Contents:

    Plan; Benefits; Preparations/Installation; Data Entry; Reports;
    Analyses; Graphics; Errors and Program Service; Reference Tools

Budget: $30,000 personnel; $7,000 printing/production

Title: Operations Manual for PRODUCTION SCHEDULER

Medium: Loose-leaf   Control #: PS-02   Priority: A

Audience: Advanced Data Entry Clerks (Y)

Responsible Documentor: Steinbeck   Completion Date: Mar 85

Sections/Contents:

    Installation; Customizing; Initializing; Data Entry; Generating Reports;
    Graphics; Errors; References

Budget: $9,000 personnel; $2,000 printing/production

Title: Developing Presentations With the PRODUCTION SCHEDULER

Medium: Spiral-bound manual Control #: PS-03 Priority: B

Audience: Art Department

Responsible Documentor: Kandinsky   Completion Date: May 85

Sections/Contents:

    Progress Graphics; Forecast Graphics; Cost Graphics

Budget: $1,500.00 personnel ; $500 printing/production

EXHIBIT 6.7   Excerpt from Documentation Set Memo

Exhibit 6.7 shows an excerpt from a completed Documentation Set
Memo, which is the output of the analysis phase and the input to the
design phase. Do not attempt to outline or write manuals until you have
thought carefully about what manuals are needed.

7

Design I: Developing
a Structured Outline

7.1   Conventional Outlines: Functions and Flaws

A conventional outline is a two-dimensional array that organizes complex material; conventional outlines show the sequence of topics and the hierarchy of subtopics. Unfortunately, conventional outlines give very little of the information needed in workplans; they do not specify the length or the scale of the sections or the document as a whole; they give no clue to the production costs of the manual. As Tables of Contents, moreover, they fail to help readers find what they need to know.

Conventional outlines organize the sequence and hierarchy of a text. To do this, they use a tiered scheme of numbers, or numbers and letters, to show subordination, and a set of topic headings to show the content or meaning of each section in the document. The typical heading contains only nouns and modifiers.

These conventional outlines are the single most common and useful tool in planning technical documents. But are they really powerful enough to perform all the design functions needed in developing structured, modular publications? Their principal benefit is to help writers to organize their own thoughts. They are the perfect planning tool for the artist working alone!

But what about other functions? Can a conventional outline help the designer of a manual estimate its length or the resources needed to prepare it? Does a conventional outline provide meaningful instructions to the several people who must write the text to go under the various headings? Does a conventional outline generate a useful Table of Contents?

When a writer works alone on a relatively small assignment, the conventional outline is often an adequate plan. But when teams of writers work on manuals, and when the documents are long and complicated, the conventional outline is not enough. It does not tell the individual authors and contributors how much to write. Neither the numbering scheme nor the typical way of writing headings (without verbs, verbals, or themes) tells the writer how long the sections should be or what they must cover.

The manager or analyst responsible for the publication gets very little data from a conventional outline. Length, and number and type of graphics – the most important predictors of cost – are nowhere mentioned. And graphics can be as much as 75 or 80% of the cost in some publications.

Ultimately, this uncommunicative workplan becomes an uncommunicative Table of Contents. And, to the extent that the reader must perceive the hierarchy in the outline, the two-dimensional outline will

---

4.0 Librarian Functions

  4.1 Librarian

  4.2 Core Image Library

    4.2.1 Cataloguing and Retrieving Programs

    4.2.2 Phases--Core Image Library

  4.3 Relocatable Library

    4.3.1 Maintenance Functions

      4.3.1.1 Cataloguing a Relocatable Module

      4.3.1.2 Library

MANAGER'S QUESTIONS:

  How Long? How Much Art? What Cost?

WRITER'S QUESTIONS:

  How Much to Write? What to Emphasize?

READER'S QUESTIONS:

  Where Do I Read? Am I Finished?

EXHIBIT 7.1   Flaws in Conventional Outlines

be useless in guiding the reader through what is really a one-dimensional product!

This last point, obviously, is a bit esoteric and needs explanation. When I say that books are one-dimensional, I am talking not about the way they are conceived but the way they are *processed by readers*. Although one paragraph may be *logically* subordinated to another, in reality it is read *after* the other. There is no actual hierarchy in a book, or even in a series of screens to be read; there is a *sequence*. Item 2 is not really below, beside, or behind Item 1; it is *after* it.

Human readers resemble most computers in that they read in sequence (not in parallel). But human readers are far less adept in assembling a sequence from a maze of loops, detours, and GOTOs. The key point is that conventional outlines, though helpful in organizing the writer's own thoughts, fail as tools for planning and designing usable user manuals.

7.2   Goals for a Structured Outline

Everyone who plans a manual—or a video program or a series of help screens—will start with a conventional outline. It will be a two-dimensional hierarchy, its items named with adjective/noun phrases. The next step, then, is to convert this outline to a structured, modular outline, in which each heading corresponds to one module of standard size and layout, and in which the language of the heading is informative and thematic—making it more a *headline*.

Most people do not have thoughts that fit into standard-sized parcels. Thus, their outlines reflect ideas that vary in length and complexity. There is no reason to believe that item 2.1 in an outline will define a section equal in length to item 2.2. And there is no way to know whether 2.2 is longer or shorter than 2.2.1.

The headings in the outline are little help. Just knowing that item 2.1 is called Administrative Aids and 2.1.1 is called Access Subsystem does not solve the problem.

For an outline to serve the design functions needed in a structured approach to user documentation, it needs two things that most conventional outlines lack:

- a style of language that specifies for the writer, reviewer, and—eventually—the reader what is actually covered in each section
- a standard that requires each entry in the outline to correspond to a certain standard-sized "chunk" of material

Now, of these two requirements, the first is far less radical than the second. Many skillful writers already use headlines rather than traditional headings—if not in their original outlines then in their tables of contents. For decades, many writers of manuals have avoided the traditional "Account Code Assignment" (three nouns) in favor of "Assigning the Account Codes" or "How to Assign Account Codes" or "Six Rules for Assigning Account Codes" or even "Why You Won't Need to Assign Account Codes." These more thematic and interesting headings (headlines) have been in the kit of more-effective writers for as long as there has been technical writing.

The other suggestion—that each entry in the outline correspond to a standard-sized item of material—is less familiar to most writers, unless they are experienced with defense and aerospace technical proposals, in which the technique is commonplace. In fact, many analysts and technical writers are astonished at the idea. How can it be possible to arrange the ebb and flow of ideas into units of uniform size? Is it feasible? Is it worth the effort?

The concept that must be grasped is that every communication is necessarily organized into standard-sized units already. Most notably, books and manuals are organized into pages of uniform size; no matter how free-flowing the ideas, they are packed in one-page chunks. Most writers and editors, though, leave this packaging up to chance. They rarely know how many pages an "idea" will take; they cannot predict the length of their "discussions." In effect, they let the typist, or the printer, or even the word processing system decide where pages will break.

Alternatively, in the structured approach to designing manuals, the goal is to design the actual object the reader/user will see: the pages or screens that will be read. So, if the book is organized by pages, why not plan and design it page-by-page? If the idea of making every section or unit the same size seems impossible, then why not make them all about the same size, with an upper limit that all must meet?

Why not convert the conventional outline to a list of specifications for each of the modules in the emerging manual or information product?

EXHIBIT 7.2   Functional Differences Between Conventional and Structured Outlines

In Conventional Outlines	In Structured Outlines
• The entries are cryptic, clear only to the author.	• The entries are substantive and informative, clear enough for review and testing.
• The entries correspond to no particular length or size of document.	• Each entry corresponds to a standard physical entity, of known length.
• The conversion to a physical product is left to editing and production of the finished draft.	• The format and layout of the physical product are inherent in the outline (provided the module has been defined).
• The scope and cost of the proposed publication are NOT apparent from the outline.	• The outline constitutes a work-plan from which costs and production requirements can be estimated easily.

7.3   Defining a Module of Documentation

A module of documentation may take many forms, as long as each module is small, cohesive enough to be independent from the other modules, and addresses a single function or theme. Each module should be *synoptic*, that is, showing the entire content of the module in a single array that will not force the user to turn pages. Modules, then, must be on one standard page, or on one odd-sized page, or on two facing pages.

A modular publication is a series of small, cohesive chunks of technical communication of predictable size, content, and appearance. Once the design – the exact sequence – of the modules is frozen, it becomes possible to treat the one, large, complicated manual as a set of many small, nearly independent manuals. Each takes only an hour or two of effort to write; each can be developed independent of the others, in any sequence. With a modular plan, it is even possible to number all the pages and figures as they are written, even though they have been written out of order!

The easiest notion of the module is *one page*. But the one-page limit, although it is perfect for some publications, is troublesome for most manuals. It is too short to present any but the most simple concepts, without crowding the page with tiny, cluttered lettering. And to merge exhibits and text on the same page calls for "copy-fitting" and other typographic skills. Furthermore, although the technology has improved, a page with art on it is a page that cannot be printed on most word processing systems.

The *two-page* module – two facing pages – is harder to maintain than the one-page, but more versatile in use. The best arrangement (developed, as already mentioned, about twenty years ago by the Hughes Aircraft Corporation) is like that shown in Exhibit 7.3.

A. Control bars, containing the document number, page, and other typical control data
B. The name of the section
C. The headline (thematic or functional heading)
D. The summary (or thesis) paragraph, which lays out the central ideas of the module
E. The full text (usually 200 to 700 words), expanding the ideas in the headline and summary
F. The exhibits – screens, diagrams, tables, drawings – on the right-hand page

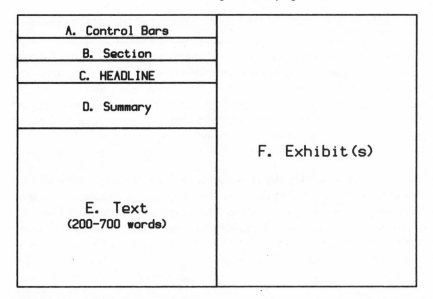

EXHIBIT 7.3   The "Hughes Module"

Note that a two-facing-page module is what people see whenever they open a publication – whether or not you intended it.

A module may fit on one page, even one 11″ × 15″ piece of computer printout. Or the arrangement of the parts in the two-page module may be moved; some documentors prefer exhibits on the left; some like to shift from right to left for the sake of variety.

Before describing some of these alternatives, however, two ideas must be stressed:

- The definition of the module, though standard and fixed, is really quite flexible, allowing great variability in the number of words or exhibits in the module; especially in the two-page module, the actual "heft" or bulk of the module can vary considerably from one ostensibly "same-sized" module to the next.
- The particular two-page format recommended has more than twenty years of successful application in all aspects of technical, industrial, government, and business communications; it is easy to learn and remarkably effective.

This basic plan can be altered as needed; two-page modules can have different arrangements of text and pictures. The right-hand page might even be a blank!

7.4   Alternative Forms of the Module, for Special Needs

The recommended two-page module is appropriate and useful for most user documentation. It is synoptic (no page turning), large enough to comprise a whole function or concept, and still small enough to be used as the unit of planning in a structured publication. There are, in addition, one-page alternatives, and even alternative forms of the two-page module that may, in some cases, better fit your needs.

The basic module contains one page. Exhibit 7.4a shows alternative arrangements of that one page and also demonstrates the mechanical drawback in the one-page module: integration of text and exhibits on a single page. If this integration is no problem, because of the form of the exhibits or the sophistication of the available production tools, the one-page module may be best.

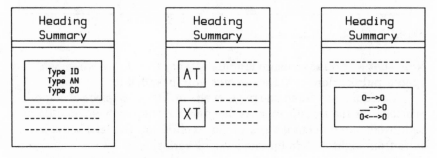

EXHIBIT 7.4a   One-Page Modules

Even when the one-page module is technically feasible, though, note that, as in software engineering, the smaller the modules, the more complicated the interfaces: the greater the number of references, loops, skips, and detours. **Very small modules do not allow you to repeat yourself, thereby forcing you to send your reader to other modules.**

If we choose the two-page module, we may also choose to realign the components in the module, as shown in Exhibit 7.4b.

Starting with the presumption that a module will be half text and half exhibits, we sometimes then discover that it makes sense in many modules to change the mix. It may be mostly text, although it is unwise to do that too often; more often, it may be almost entirely charts, listings, or some other exhibit. We may also shift the exhibits to the left, or even expand the module with a foldout.

Exhibit 7.4c shows the possibilities of odd-sized pages: half-size pages, or smaller, for very simple machines; 11″ × 15″ printout pages

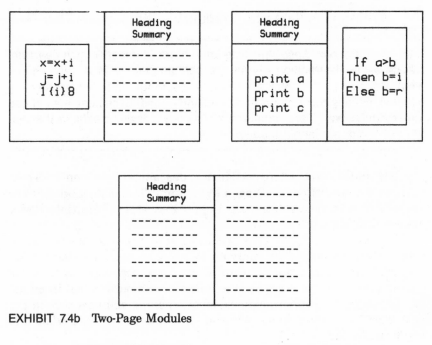

EXHIBIT 7.4b   Two-Page Modules

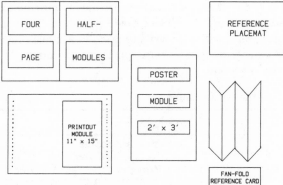

EXHIBIT 7.4c   Odd-Sized Modules

(24 rows by 120 columns); $2' \times 3'$ posters; reference placemats; fanfold reference cards.

A module may be of any workable size or shape, as long as it is big enough to communicate a whole concept, small enough to be easily specified in a plan, and synoptic.

Thus, the perfect module might be a single screen – preferably one that could be read without scrolling. In that case, the limits and size of the module would be constrained by the limits of the developer's video display.

7.5   Writing Headlines for Modules

Each module, regardless of its size and shape, is impelled by a *headline*. Headlines, unlike traditional headings, contain themes, ideas, assertions, even arguments. In contrast, traditional headings, even when they are detailed, usually contain only modifiers and nouns. Each headline defines "one module's worth" of documentation: a body of text and exhibits that can be presented in a single module.

The traditional heading used in outlines and tables of contents contains only nouns ("Logon," "Logon Procedure," "Power Redundancy") or nouns and modifiers ("Alternate Logon Procedure," "Multiple Power Source Redundancy").

These traditional headings give little clue to the actual scope or intent of the section, and certainly no clue to its length. Neither the author who must write the section nor the reader who is searching the table of contents really knows what the writer of the outline had in mind.

In contrast, headlines express ideas, themes, emphases. Both the author and the reader know why the section is written and what it is supposed to do.

The real key to writing an effective outline is **knowing exactly what you want said in the module.** Notice how I put that: not *what you want to say* but *what you want said*; writing good headlines may be the first step to getting other people to help you write the manuals you do not choose to write all alone.

EXHIBIT 7.5a   Types of Headline

Style	Example
VERBALS	Opening an Account How to Open an Account Two Ways to Open an Account
ADVERBIALS	Where Signature Cards Are Filed When Accounts May Be Opened Why XYZ Accounts Are No Longer Opened
APPOSITIVES	New Accounts: Who May Authorize One? Account Codes: How to Interpret Them Form 99: The Need for ID Verification
INJUNCTIONS	The Need for New Account Classes The Importance of Updating Subscriber Lists The Danger of Unverified Transfers
SENTENCES	Ten Accounts Can Be Processed Each Hour Memorize the Code Formula Does Everyone Get a Monthly Statement?

Note that the knack of writing headlines is, in another sense, independent of the knack of organizing a manual into standard-sized modules. Clearly, one can write headlines without any regard for the heft or length of the material to be covered under them. In fact, an alternative way to develop a structured outline is to go through the intermediate stage of a "substantive outline," one in which the headings have been recast into headline style without regard for modules. In this alternative (discussed later), the substantive outline is then further refined into the structured outline.

Although it may seem somewhat confusing to solve both problems at once – thematic language and module-sized chunks – many writers find it easier to plan this way. Knowing the size of a module tends to refine the headline, making it sharper and clearer.

Exhibit 7.5b shows some traditional headings taken from my own library of user manuals and also what their authors seemed to have meant by them. (I have left a few blank, just in case you want to try your hand at a hypothetical headline or two.)

EXHIBIT 7.5b   Converted Headings

Before	After
Access Methods	Two Forms of Access: Sequential and Direct
Files	How the System Verifies File Names and Definitions
Executive Libraries	The Time-Sharing System Provides Standard and Specialized Processing Routines
Program Debug	How to Use the DEBUG Facility
FDEBUG Example	Example: Debugging a FORTRAN Program with FDEBUG
Namelist Facility	
LOAD and USE Commands	
Transparent Write	

7.6   Demonstration: Headings into Headlines

The exhibits in this section show how traditional outlines look when they are converted to structured outlines, lists of headlines. Note that the tone of the headlines may be either light and conversational (a kind of "marketing" style), or, alternatively, straightforward and technical.

In the first example, Exhibit 7.6a, the outline for an Installation Plan – one of the essential user documents – is shown in both forms. The "before" version is typical of the style of DP departments, but the "after" version is vastly more likely to communicate clearly with the user departments and others affected by the installation.

Exhibit 7.6b shows the outline for a typical user guide to a typical business package. Note, however, that there are two versions of the re-

---

EXHIBIT 7.6a   Outlines for an Installation Plan

Before:
1. Site Preparation
    1.1 Electrical Requirements
    1.2 Physical Requirements
2. Assembly
    2.1 Attachments
    2.2 Interfaces
3. Communications
    3.1 Communication Protocols
    3.2 Alternative Configurations
4. Testing
    4.1 Communications Test
    4.2 Mechanical Test
    4.3 Software Test

After:
1. Installing the Necessary Electrical Fixtures
2. Ensuring the Right Temperature and Cleanliness
3. Attaching the Cover and Paper Feeder
4. Choosing and Attaching the Right Cables and Connectors
5. Connecting the Plotter to the Computer
6. Setting the Communication and Protocol Switches
7. Running the Communications-Check Program
8. Diagnosing Start-Up Problems
9. Solving Communication Problems
10. Solving Mechanical Problems
11. Setting the Switches for Your Graphics Software
12. Testing the System with Your Software

---

vised outline: one showing the appropriate language, scope, and sequence for an executive in the financial department, the other showing a financial clerk how to operate the system.

Notice how traditional outlines do almost nothing to help the writer anticipate the audiences and functions of the publication.

EXHIBIT 7.6b   Outlines for a Business Product

Before:
1. Introduction
2. Accounting Highlights
   2.1 Account Structures and Levels
   2.2 IDs and Descriptions
3. Systems Supported
   3.1 Vendor Payment
   3.2 Budget Development
   3.3 Budget Control
   3.4 Financial Reporting
4. Appendix: Sample Outputs

After: Executive Version
1. Using a Financial Information System
2. Defining Accounting Codes to Meet Legal Requirements
3. Defining Accounting Codes to Support Planning and Analysis
4. Designing Financial Reports
5. Analyzing Current Patterns of Expenditure
6. Simulating Alternative Budgets
7. Enforcing a Budget

After: Clerical Version
1. Entering Data
   1.1 Entering a Receivable
   1.2 Entering a Receipt
   1.3 Entering a Payable
   1.4 Entering a Payment
   1.5 Entering a Budget
   1.6 Editing an Incorrect Entry
2. Getting Reports
   2.1 Running the Monthly Financial Report
   2.2 Running the Quarterly Financial Report
   2.3 Running the Year-End Financial Report
   2.4 Running the Annual P&L Statement
   2.5 Running the Budget Comparison Reports
3. Appendix: How to Respond to Error Messages

7.7   How to Convert from One Outline to the Other

Almost no one who uses the modular approach to publications begins directly with a structured outline of headlines. Rather, the typical way to develop a list of headlines is to prepare a conventional outline and, then, "decompose" it into a series of headlines—each representing about one module's worth of material.

Converting a conventional outline, then, usually involves two simultaneous changes: Converting the language of the headings into the style and syntax of headlines; and writing headlines that generate (as best as can be predicted at this point) one module of material, however "module" has been defined.

The first of these two tasks is the easier. Ironically, most people, when they plan a report or manual, think in headlines before they write traditional headings. The writer knows that he or she wants to explain "An Easier Way to Correct an Error in an Old Account"; but the outline contains the entry: "Alternative Old Account Error Correction Procedure." Or another writer wants to demonstrate "Five Ways That System B Eliminates Transmission Delays"; but the outline contains the entry: "System B Transmission Program." So, in some ways, writing headlines consists in reverting to an earlier conception of what was to be said.

The other part of the task is difficult; most people, especially professional technical writers, are unaccustomed to outlining in uniform chunks. Although they may see the communication advantages in using headlines—indeed, many good writers already use them, some without realizing that they are doing anything remarkable—they are uncomfortable with answering the question: Will "An Easier Way to Correct an Error" fit into one module, especially when they are told that it is not consistent with the technique to call one of the modules "An Easier Way to Correct an Error—Part 2."

As a result, some organizations and writers choose to add an intermediate step, the previously mentioned "substantive outline." In this, the headings need not generate uniform quantities of material—as long as they are clear and evocative enough to support a productive review of the planned document. After a while, though, they usually discover that this is an unnecessary step, that it is relatively easy to perform both conversions at once, and that, even if a structured outline contains many incorrect predictions as to the number of modules, it is still far more interesting and useful than the substantive outline alone.

Remember that, as of this stage in planning the manual, there is no need to be absolutely sure that each headline corresponds to exactly one module. In time, documentors learn to write headlines with the right

"heft." But there is no need to be anxious about the accuracy of the estimate. The next design steps will reveal whether the headlines merit one
module, or more, or fewer.

EXHIBIT 7.7   Converting a Conventional Outline

Conventional Outline	Structured Outline
I. Introduction	
A. Background —————————	1. Problems in Batch Processing of Retail Transactions
B. Project History—————————	2. A Three-Stage Conversion to TRANSACTIONS
II. Operating Highlights	
A. Transparency —————————	3. How the New System Simplifies Operation
	4. Elimination of Batch Activities
B. Transaction-Based—————————	5. Each Transaction Updates *All* the Files
	6. Data Are Typed Only Once
C. Security —————————	7. Access Is Strictly Controlled
III. Functions	
A. Retail—————————	8. Recording Retail Transactions
	9. Opening Retail Accounts
B. Accounting—————————	10. How the Retail Data Reach the Accounting Files
C. Inventory Control —————————	11. How the Retail Data Reach the Inventory System

Exhibit 7.7 shows part of a conversion from traditional headings
to headlines; no one can be sure whether each headline is one module's
worth, but, at this point, no one has to be completely sure.

Note that the sequence of the new headlines corresponds to the
original sequence of headings; the headlines are a subset of the headings,
in the same sequence. Occasionally though, the order of the parts in the
original outline needs to be changed. When designers try to eliminate
loops and GOTOs from their publications, they also sometimes decide
to change traditional sequences. For example, there is rarely a reason
to begin a user manual with background material. Usually, the appropriate topic is benefits and advantages.

7.8   Is It Possible to Predict the Number of Modules?

People who do not write much are skeptical of the claim that, at this early stage, it is possible to estimate the size of the modules, to be sure that what is defined as a module will fit into the space provided. Even professional technical writers, who often make rough estimates of the length of a piece by looking at the outline, are reluctant to predict length so precisely. Actually, it takes less than a month for most people to learn to estimate the heft of the modules and, thus, the number.

Obviously, it is impossible to predict from an outline the *precise length* of any of the items in that outline. Fortunately, though, devising a structured outline does not require such an estimate.

In this scheme, a module is an *upper limit* (one or two pages, of any dimensions), not a uniform size or length. It is far easier to make sure that no module is too large than to make sure all modules are identically long.

Note also that modules vary considerably in length and content, even if they all fit into the same two-page spreads. By adjustments of artwork and typography, a module might have as few as 200 words, with or without an attendant increase in the size or number of exhibits, or as many as 1200 words.

Furthermore, the structured outline is not the last chance to estimate the size of the modules. Later, when the outline is finished, the designers will write a small spec for each module, at which point they may decide that what they thought was one module is, in fact, more than one. And, still later, when all the specs are mounted in a model or storyboard, there is one more chance to revise the estimate.

Generally, it takes writers only a few weeks to develop a sense of the module-sized chunk of material. Amateur writers often learn the technique quicker than professionals, who need a few days to unlearn some of their old habits. Although the idea of setting a standard physical limit on the size of a concept or procedure seems harsh and restrictive at first, in a short time it reveals itself to be a useful discipline that encourages intellectual creativity.

The only change in publication policy needed to implement modular publications successfully is a willingness to allow some white space in manuals – the consequence of short modules. For some publications managers, the sight of unused white paper is anathema. They see expense and waste; they do not see the increased readability and maintainability of the publication – which sometimes saves thousands of times as much as the "less wasteful" printing could have saved.

To the question "Won't there be a lot of blank space in a modular publication?" the answer is "Probably."

8

Design II: Developing
a Storyboard and Model

8.1   The Value of Models in Solving Documentation Problems

Models save money and effort. They allow us to experiment and innovate without risk. They enable us to test ideas and products with slight expense — and to correct faults at a fraction of what it costs to repair a production version of the product. Without models, documents are unlikely to be tested as hard as they should be, and even less likely to have their bugs corrected.

Exhibit 8.1 depicts one of the most important functions in the world of work: the relationship between the cost of correcting an error and the time at which the correction is made. The function is exponential; that is, the curve not only accelerates, it accelerates at an ever faster rate.

The more complicated the project, or the more unfamiliar and risky the technology involved in the project, the faster the curve accelerates. That is why very simple and familiar projects do not need as much analysis and design — although sometimes even the most familiar task can be radically improved by testing old assumptions.

A *model* is a representation of one thing by another. Models are made either from different materials (clay instead of steel, paper instead of switches) or on a different scale (a miniature of a building, an oversized model of an atom). The materials and scale of the model make it easier to build and, more important, easier to change.

Effort, Cost

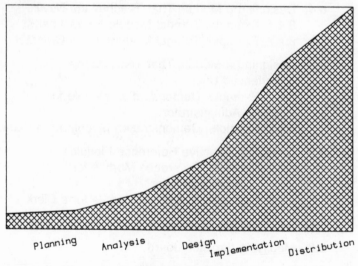

Planning   Analysis   Design   Implementation   Distribution

Project Phase

EXHIBIT 8.1   Exponential Increase in Cost of Making a Change or Correction

Models are relevant to writers of user documentation in two important ways. First, user documents and operations guides can be models themselves. In the most sophisticated development groups, the design team will write an operations guide as a way of specifying and testing the "operator interface." In other words, the developers define the human component of the system by writing a friendly, before-the-fact user publication, which then guides the subsequent design of the system or program.

(User documentation written at the beginning of product development is rare and exceptional, but clearly—as a growing number of firms has discovered—the best way to work. Among other benefits, writing early user manuals forces developers to think about the thing they usually leave for last: what the *people* will do with the system.)

The second relevance of models is in the development of the manuals themselves. Manuals need models. Before writing and drawing a draft of any publication longer than a few pages, documentors should devise a model of the publication that makes clear what will happen within each module and that also shows all the links and couplings across modules. In effect, it should be possible to evaluate the accuracy of the technical content in each module and predict the number of loops and branches across modules. As in software engineering, **the greater the number of possible paths through the document, the less reliable and more error-prone the process of reading it**.

Models are for testing. And the purpose of testing is to find flaws, mistakes, bugs. Models and tests make vivid our misunderstandings, focus our disagreements, underscore our schedule and cost problems, prove that we cannot have what we want, or that we cannot have everything we want in time. In short, **models force us to recognize our errors and redo our work**.

And that is why few writers, and nearly as few DP and MIS people, want to use them. Most of the clients I meet do not want to know what is wrong with their work. They do not want to be reviewed, tested, inspected, verified, validated, evaluated, or "walked through."

Surely, no one likes criticism. But the longer people work on a manual or system, the less receptive they are to critical opinions. An added benefit of working with models, therefore, is that they enable people to see the flaws in their plans—before they have fallen in love with them!

8.2   Building a Model for Each Module

The Module Spec is the detailed plan for a module, written collaboratively by the technical expert on the documentation team and the communicator. Each Spec includes the headline (transcribed from the outline), a one-paragraph summary of the module, an indication of the exhibit, and, if necessary, some additional notes for the writer.

The *structured outline* – list of headlines – is a far clearer design of the emerging manual than the traditional outline. But it is still only the preliminary design. At this stage, the people preparing the outline cannot be sure that they have written the best headlines, that the proposed modules are in the right sequence, that the needed material is in, or that the unneeded material is out. To be sure, they cannot really know whether

Structured Outline

1. How to Start the Computer

2. How to Copy the Distribution Diskette

   2.1 Copying with ONE Disk Drive

   2.2 Copying with TWO Disk Drives

3. How to Name Your File

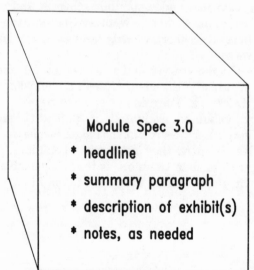

Module Spec 3.0
* headline
* summary paragraph
* description of exhibit(s)
* notes, as needed

EXHIBIT 8.2   Writing a Spec for Each Module

the material they plan to cover in each module can be fit into the space allowed.

To build a true model of the book, detailed enough to review and test before the draft is written, there should be a plan for each module. As Exhibit 8.2 shows, each headline gets a plan of its own: a Module Spec.

The kind of paper (or form) used in writing Specs varies with the size and content of the Module. For the two-page spread, though, the most practicable (and accessible) form is the blank side of 11″ × 15″ computer paper – even the backs of old printouts. Each 11″ × 15″ sheet is slightly smaller than the 11″ × 17″ module being planned.

To prepare the Module Spec, the designers

1. *transcribe the headline* to the Module Spec.

2. *write a one-to-four sentence summary paragraph for the module.* This paragraph will actually appear in the finished publication; it should be a self-sufficient paragraph that does NOT contain remarks like "This section describes..." or "The procedure below ..." Rather, it should be a précis version of the whole module. The summary may, however, refer to the exhibits in the module. In the terminology used for scientific papers, it is an informative abstract rather than indicative abstract.

3. *describe the exhibit.* The spec must contain just enough information so that the reviewers of the plan can know exactly what the exhibit will be; or the spec may contain a rough sketch, approximate graph or table, even just a few key phrases or parameters that define a chart or table. Ideally, there should be just enough description so that someone else could complete the details and produce the exhibit envisioned by the designers.

4. *add notes, as needed.* Just in case the headline, summary, and exhibit are not clear enough – as they usually are – the designers may also add a few notes describing what the module will contain. Again, just enough so that someone who had reviewed the spec would not be surprised by the finished module.

After a brief "learning curve" (two or three hours), most people find that they can write a Module Spec in 10 to 15 minutes. And in some publications, in which a series of modules falls into the same pattern, the repetitive Module Specs can be prepared in only about 5 minutes each.

8.2   Building a Model for Each Module
8.2.1   Does Every Module Need an Exhibit?

Probably, every module in a user manual can benefit from an exhibit—a diagram, a screen or two, a drawing, a table of words, or a table of numbers. In a well-designed module the exhibit is *redundant* with the text, not supplemental to it.

Presume that every module will have some exhibit. That is, plan on having at least one exhibit in each module, but be prepared to abandon the idea if, after hard thought, you cannot think of one. Or if there is not enough space.

The material communicated in the exhibit overlaps—in some cases duplicates—the material in the text. In fact, in the best module there is double repetition; the headline and summary state the content, which is echoed in the exhibit, which is further echoed and enhanced in the detailed text. Does not all this redundancy violate some principle of concise, technical communication?

No. Redundancy violates a principle of efficiency; redundancy raises the short-term costs. Indeed, leaving out graphics altogether also reduces short-term costs. **Remember that redundancy is absolutely necessary to ensure effective communication.** And redundancy of pictures and text is the shrewdest way to communicate technical information to audiences with different "learning styles."

Most of the exhibits will fall into these main categories:

- *Flow and process diagrams*, abstract symbols that represent either the physical movement of events and material or the logical movement of data and ideas. There are also diagrams that clarify procedures.
- *Displays and screens*, duplications, reproductions or renderings of what actually appears on the video display or other input/output device; one or more screens per module is the most typical method of documenting on-line systems.
- *Drawings and representations*, any attempt by drawing or photography to depict actual objects or, occasionally, people; technical drawings are usually the preferred method, because they are easier to reproduce, but photographs are used when the emphasis is on reality or credibility.
- *"Verbal" graphics*, exhibits made up mainly of words, with some simple embellishments, boxes, and arrows; although rare in technical or user manuals, verbal graphics can be especially useful

EXHIBIT 8.2.1   Exhibits for User Documents

DIAGRAMS	• Flowcharts • Networks • Data Flow Diagrams • Structure Charts/HIPOs
DISPLAYS	• Screens • Worksheets, Forms • Windows, Panels
PICTURES	• Illustrations • Photographs • Design Graphics (Projections)
VERBALS	• Word Tables • Pseudocode or "Structured English" • Decision Tables/Trees • "Information Maps" • Listings, Programs • Playscripts
MATHEMATICS	• Statistical Plots • Pie/Bar/Line/Surface Charts • Equations, Models

in plans, briefings, and training materials. (The module you are reading now has a "verbal" graphic.)

- *Playscript/dialogue,* techniques that show operators, users, and equipment as though they were participants in a play; playscript, a set of techniques developed originally for manual systems and procedures, is extremely adaptable for data preparation procedures and interactive procedures.
- *Mathematical and statistical exhibits,* mathematical expressions, graphs, statistical tables, and the full range of exhibits associated with science and engineering.

Can the same exhibit appear in more than one module? Of course. Although many technical editors and publications managers will resist the suggestion, I urge you to repeat an exhibit rather than committing that most serious error: referring to an exhibit that cannot be seen.

In practice, however, most documentors discover that each of the several references to the "same" exhibit are, in fact, references to different fields on the screen; different cells in the table. Although the usual practice is to produce the exhibit once and refer to it from several places in the text, the smarter policy is to create separate exhibits for each reference, or, in some cases, the "same" exhibit with different parts emphasized or highlighted.

8.2 Building a Model for Each Module
8.2.2   What If There Is a Little Too Much for One Module?

By adjusting the size and placement of exhibits, or by changing the type style or page format, most authors find it easy to "squeeze" a bit more information into a tightly packed module. Generally, if an idea or procedure is too big for one module, though, it needs at least three.

When documentors try to write in standard-sized modules, one of their first concerns is the problem of the module that is just a little too big to fit in the one- or two-page limit.

Recall, however, that the standard module is an *upper limit*, not a uniform size. When designers and writers start to think in modules, they are really thinking in chunks of material that are *up to* the size of one module – not in chunks of uniform size. As a result, most modules will have white space in them, which can only make them easier to read.

For those modules that are bursting at the seams, though, there are numerous remedies. Artwork can be shrunk; even text can be reduced somewhat, although most publication professionals prefer not to change type sizes from page to page.

In addition, there are many ways to expand the capacity of a module – without producing clutter or making it harder to read. Text presented in *columns* usually allows the writer to fit 10 to 20% more material into a space without deleterious effects. Text typed with *proportional printing* yields a similar benefit. Note, however, that the typical office word processor these days does *not* use proportional spacing and, further, justifies the right-hand margin by adding extra spaces between words. Simply turning off the right-hand justification and using "ragged right" printing will increase the per-page capacity of many word processing systems.

There is even the option of *removing* some material from the fat module, provided one is sure that its loss will not interfere with the clarity and effectiveness of the module.

That a process or concept is too large for one module of documentation is a powerful piece of test data. In almost every case, it means that the item is too big to be regarded as one entity. Especially when the module is the spacious expanse of the two-page spread of 8½ by 11″ pages, an entity that will not fit the module is probably best regarded not as one thing but as a small organization of things.

As Exhibit 8.2.2 shows, breaking a long idea into Phase I and Phase II is less coherent and intelligible than beginning with an overview or top-level view which explains that the process has two phases, and which

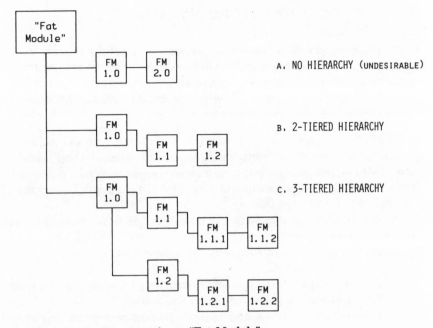

EXHIBIT 8.2.2   Decomposing a "Fat Module"

then offers a module for each phase. So, a process with two phases needs three modules, with three phases four, and so forth.

In some outlines, the headlines are so broad that it takes a three-level hierarchy to present what we first thought was only one module (as the exhibit illustrates).

Operations and procedures that can be presented within a single module are more easily learned and followed than those needing hierarchies and branches. Often, then, the wisest thing for a documentor to do – having discovered an especially fat module – is to persuade the developer to change the procedure itself. This will not only simplify the writing of the manual: it will make the technology more usable and reliable.

8.3   Designing a Module That Motivates

A motivational module is one whose purpose is to get the readers to do something they do not want to do. It must convince the readers that they will *benefit* from the process or technique recommended in the module, that they will gain more from doing what is proposed than from not doing it.

Even though documentors may think of themselves as "technical people," they nevertheless must sell ideas and methods to their readers. Operator manuals and user manuals almost always contain some motivational material. That is, modules that convert the *features* of the system into *benefits* for the reader.

Every system replaces some other system; the differences between the former and the latter are the features to be described.

Most features fit into relatively few categories:

- *Physical aspects* — components, size, weight, temperature, location, quantities, general appearance, sound
- *Operating aspects* — speed, cycle rate, number of steps, "capabilities" (what it will or will not do), compatibility with other things
- *Accessibility* — quantities on hand, learning time, delivery time, service time, acquisition costs, operating costs
- *Performance features* — elegance, rigor, accuracy, precision, reliability, versatility, expandability

To repeat, any system or procedure you recommend must differ in some of these characteristics from the one you wish to supplant or replace. And the problem is to map one or more of these features onto one or more of the benefits.

The most common mistake is the Skill Trap. That is, many writers think there are a great many people who find several of the features above inherently desirable and worthwhile. There are fewer of these people, however, than engineers and analysts believe.

The most common motivator, of course, is Material Benefit. The recommended procedure will help the users to *make money or save money*. One of the hardest sells, of course, is to convince people that high short-term costs will be repaid with higher long-term savings.

In addition to material benefits are Psychological Benefits: respect, status, prestige, recognition, affection, relaxation, comfort, power, independence. People who specialize in motivation claim that these benefits count for more than small material advantages — especially in communicating with junior employees. Power, for example, is attractive to the executive who wants more control over his or her organization; but

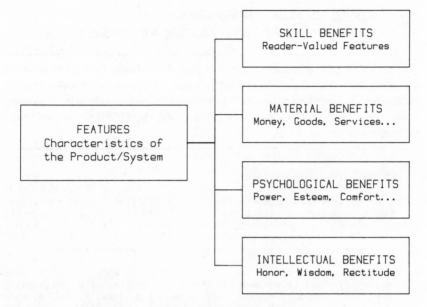

EXHIBIT 8.3 Designing a Module That Motivates

it is also attractive to the clerk who wants more control over his or her "free time."

Less often used in American business and government is the appeal to Moral and Intellectual Benefits: doing something because it is right, or just, or enlightening, or ennobling. In some specialized institutions (like universities or religious organizations) and even in some entire cultures, these appeals persuade people to try new activities as eagerly as people in our society are drawn to the "better bottom line."

The point to be stressed is that none of these benefits is obvious in the features. Even the *cost features* of a new system may need extensive explanation and justification to prove that they provide a material benefit to the reader.

The documentor must analyze what the readers/users want and must show explicitly – in the summary paragraph of the module – how the recommended action can get it for them. And the exhibit should, in most cases, show the comparative advantages of the two approaches side-by-side.

8.3   Designing a Module That Motivates
8.3.1   Example: Motivational Module for an Executive

The example below is from a system plan for a large international bank (not written by me). See how the simple table in the exhibit proves that the recommended product saves money, reduces stress, and builds on existing skills. The summary paragraph would have been better if it developed these motivators more explicitly.

---

3. CRX Versus the Alternatives

There are three alternatives to adopting the CRX Loan System: continuing to develop current commercial loan (C/L) products; examine new C/L systems; abandoning certain business functions. All the alternatives are either too slow or too expensive.

The CRX Loan System is similar to, and compatible with, systems currently used elsewhere in the bank. Because its programming language, record layouts, and input/output formats are already familiar in other divisions, it will be easy and efficient to merge CRX with the current loan procedure. Note the following implementation advantages:

- The bank's systems analysts already know the product.

- Many users have tested the product.

- Forms, procedures, and documents are already developed.

- The software is already paid for.

The Alternatives

If we continue to develop current C/L products... there will have to be major changes in the current version 2.4. We will have to add new fields to every screen and to expand the masterfile. This change would be so slow and expensive that new loan products could become outdated before they are implemented.

If we examine other C/L systems... we will lose time in the evaluation. Any new system would have to be purchased (whereas we already own CRX) and still more time would be lost in training and orientation. [Also, preliminary assessment shows that CRX is better than its competitors.]

If we abandon certain functions... we will be unable to book many high yield business loans. Further, we will have to support an increasingly fragmented and inefficient set of lending organizations.

CRX Versus Alternatives

CRX	Alternatives
* small cost	* expensive
* fast development possible	* slow development
* already tested and proved	* start from scratch
* system expertise available	* ???

EXHIBIT 8.3.1

8.3   Designing a Module That Motivates
8.3.2   Example: Motivational Module for an Engineer

The example below is from a user guide aimed at engineers in a military installation (not written by me). The motivators are signaled clearly in the summary paragraph: not only skill, but *power*—freedom to work the way the engineer chooses to work. The exhibit compares the skill features but does not reinforce the power motivation.

```
1.3  RANDOM ACCESS CURVE FILE ADVANTAGES FOR ENGINEERS
```

In the past, we were stuck with a rather inflexible curve file design. Now, with random curve files, you have automatic units conversion, meaningful curve and parameter names, and much more varied support software. Curves can now have up to eight independent and 50 dependent parameters.

In the past, curves were identified by relatively meaningless "curve numbers" and the data contained within them was referred to as the X, Y, Z, or W variable of the curve. Woe be unto the engineer, whose cohort made the CLALFA curve with ALFA in degrees, while his program assumed ALFA would come back in radians. Why did the programmer hit the roof when you wanted your curve plotted with lines of the "W" variable, instead of lines of "Z"? Or, how about the time your UFTAS printout was lost in a sea of extrapolation messages?

Now, curves are identified by names (hence, no more error prone connection to a range of curve numbers). Variables making up the curve also have names, as well as units. Now you can tell the software what units you want your data to be in and the new automatic units conversion aspect takes care of the rest.

The plot package allows you to select any variable as the x-axis and any other variable to have plotted as "lines of". You may select what units you want your data plotted in and you may even select what names your variables are called on the plots.

Curves can now be much larger than in the past. Curves that are a function of up to eight different independent variables can now be handled. Curves now can have multiple dependent parameters. In fact, each curve can have up to 50 dependent parameters.

Besides the plot package, the support software includes creation, audit, merge, print, reformat of UFTAS curves, and Tektronix digitization programs, making up a complete and powerful package.

ADVANTAGES OF NEW
RANDOM ACCESS
CURVE FILES

CAPABILITY	OLD CURVE FILE	NEW RANDOM CURVE FILE
1. CURVES REFERENCED BY	NUMBERS	MEANINGFUL NAMES
2. PARAMETERS REFERENCED BY	X,Y,Z,W PARAMETERS	MEANINGFUL NAMES
3. PARAMETERS UNITS	NO	YES
4. AUTOMATIC UNIT CONVERSION	NO	YES
5. PLOT FLEXIBILITY	NONE	MUCH MORE FLEXIBLE & VARIED
6. NUMBER OF INDEPENDENTS	3 MAX.	8 MAX.
7. NUMBER OF DEPENDENTS	1 ONLY	50 MAX.
8. LOOK-UP ROUTINES CALLING ARGUMENTS	STANDARD	MORE FLEXIBLE & MEANINGFUL
9. SUPPORT SOFTWARE	CREATE PLOT PRINT	CREATE PLOT PRINT AUDIT MERGE CONVERSION DIGITIZING
10. EXTRAPOLATION	MANDATORY	CAN BE TURNED ON OR OFF
11. EFFICIENCY	SLOW	FASTER

EXHIBIT 8.3.2

8.4   Designing a Module That Teaches the Novice

A tutorial module must present a single concept or task and then test the reader to see if the concept or task has been learned. Documentors define the aim of the module in terms of a particular item to be mastered and then require the reader to prove mastery: by answering a question, completing a simple operation, or advancing along a progression of tutorial instructions.

A tutorial module contains **one new thing**.

Before writers can design an effective tutorial module, they must be able to say exactly what they want the reader to learn from it. And the most useful way to describe that objective is to think of some task or test, keyed exactly to the concept or idea being taught. In effect, if the reader can answer a certain question, make a certain choice, finish a certain process, or otherwise prove mastery of the concept, then the module will have been effective. In more-sophisticated teaching materials, one may even specify other limiting conditions, such as how much time is allowed for the task, or how many wrong answers are permitted among the right answers.

The sample in Exhibit 8.4a might look painfully obvious to an experienced operator, but it frequently is just the right way to communicate with a novice. (Note: Tutorial modules frequently take very little room; it is not uncommon to present them in one-page modules, or even in pages that are smaller than the conventional 8½" × 11".)

The sample in Exhibit 8.4b is more typical of operator materials. Naturally, since the task is to generate the "solution screen," the manual must be used at a live terminal. (It is difficult to imagine a way to present a series of tutorial modules without having the reader at a working system. Especially for the inexperienced reader, it is nearly impossible to learn basic tasks without actually doing them.)

The sample multiple-choice question in Exhibit 8.4c could appear

PRESSING <L> TELLS THE SYSTEM
TO CHANGE THE *LOGGED DISK DRIVE.*

WHAT DOES THE "L" STAND FOR?

EXHIBIT 8.4a

Using the CURSOR CONTROLS and the INSERT key,
Turn Screen # 1 into Screen # 2.

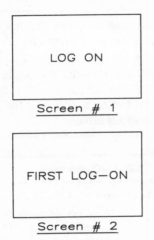

LOG ON

Screen # 1

FIRST LOG—ON

Screen # 2

EXHIBIT 8.4b

If you format the diskette with the command FORMAT/S,
then the programs you save on that diskette will be——

F1 Copy—Protected
F2 Bootable
F3 Compatible with all S—Type Computers
F4 Automatically backed—up

EXHIBIT 8.4c

either in a book (probably a programmed textbook) or, better, as part
of a computer-assisted series of instructional screens. Interestingly, although a programmed textbook is probably the most effective way to
teach a novice user, the best programmed texts are designed to force
the reader to skip, jump, branch, and detour! The problem is that, if insecure or inexperienced readers get lost in a programmed text, they never
find their way back. And the solution is the *on-line tutorial*, a programmed
text with "transparent" branches.

Obviously, more-complicated training materials call for the skills of
a specialist, an instructional technologist.

8.4   Designing a Module That Teaches the Novice
8.4.1   Example: Tutorial Module for an Analyst

The excerpt below is from a teaching manual intended to introduce financial analysts in a consulting firm to a new program that creates and manipulates statistical tables. (It was not written by me.) In this case the reader is a novice to this particular system — not to data processing in general. The module lacks a summary paragraph, and it would be improved by the addition of a small question or two to assure that the reader can distinguish among the three ways to store tables.

---

18. Three Ways to Store TABLES Commands

You may choose one of three ways to store TABLES commands.

A TABLE MEMBER	is a set of TABLES commands stored as a member of a partitioned TABLE FILE. You may store only DEFINE, IDENTIFY LIST, and ADD TRANSFORMATION commands in a TABLE MEMBER.
A BINARY TABLE	is a set of TABLES commands similar to a TABLE MEMBER, but encoded in a compact, binary format for quicker and more efficient storage and retrieval. A BINARY TABLE may contain only DEFINE, IDENTIFY LIST, and ADD TRANSFORMATION commands. You cannot alter or edit a BINARY TABLE.
A COMMANDS DATA SET	is a set of TABLES commands, stored as card-images, but not as part of a TABLE FILE. A COMMANDS DATA SET may contain any TABLES commands. You may edit or alter a COMMANDS DATA SET.

To create a Table Member or a Commands Data Set, use a text editing program such as UNI-COLL's QED or IBM's EDIT. To create a Table File in which to store Table Members, use BUILD TABLE FILE. To create a Binary Table, use WRITE BINARY TABLE.

To recall a Table Member, use READ TABLE, after you have SELECTed the Table File to which the member belongs with SELECT TABLE FILE. To recall a Commands Data Set, use READ COMMANDS DATA SET. To recall a Binary Table, use READ BINARY TABLE.

Three Ways to Store TABLES Commands

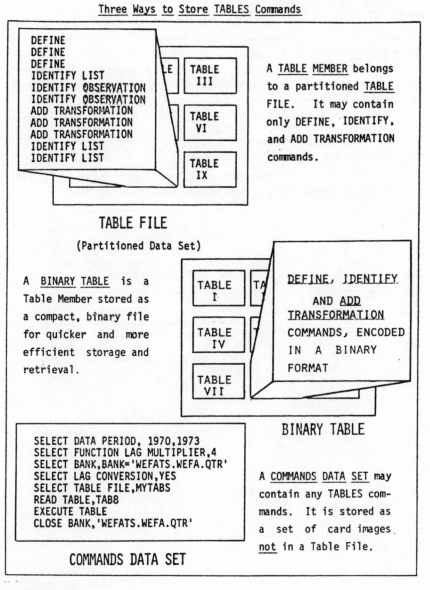

DEFINE
DEFINE
DEFINE
IDENTIFY LIST
IDENTIFY OBSERVATION
IDENTIFY OBSERVATION
ADD TRANSFORMATION
ADD TRANSFORMATION
ADD TRANSFORMATION
IDENTIFY LIST
IDENTIFY LIST

TABLE III
TABLE VI
TABLE IX

A TABLE MEMBER belongs to a partitioned TABLE FILE. It may contain only DEFINE, IDENTIFY, and ADD TRANSFORMATION commands.

TABLE FILE

(Partitioned Data Set)

A BINARY TABLE is a Table Member stored as a compact, binary file for quicker and more efficient storage and retrieval.

TABLE I
TABLE IV
TABLE VII

DEFINE, IDENTIFY
AND ADD
TRANSFORMATION
COMMANDS, ENCODED
IN A BINARY
FORMAT

BINARY TABLE

SELECT DATA PERIOD, 1970,1973
SELECT FUNCTION LAG MULTIPLIER,4
SELECT BANK,BANK='WEFATS.WEFA.QTR'
SELECT LAG CONVERSION,YES
SELECT TABLE FILE,MYTABS
READ TABLE,TAB8
EXECUTE TABLE
CLOSE BANK,'WEFATS.WEFA.QTR'

A COMMANDS DATA SET may contain any TABLES commands. It is stored as a set of card images not in a Table File.

COMMANDS DATA SET

EXHIBIT 8.4.1

8.4   Designing a Module That Teaches the Novice
8.4.2   Example: Tutorial Module for an Operator

The excerpt below is from a user manual that teaches a procedure for setting up telephone links on military bases. (It was not written by me.) It teaches the operator one new skill—use of the LOAD command. The module would be improved by the addition of a small test question.

Module 11.0   Users/Operators Manual   Jan 84

Use the <<LOAD>> Option to Move Data from "Post" to "Base"

You may use the <<LOAD>> option from either <<TERMINALS PROCESSING>> or from <<REQUIREMENTS PROCESSING>>. When you call for <<LOAD>>, the system reads the data from the auxiliary data base "POST" and, then, defines requirements, terminals, and buildings in the "BASE" data base.

After you have followed the procedures described in Modules 10 and 10A, you may invoke the <<LOAD>> option in the <<FILE PROCESSING>> part of the system. Whenever you call for this option, the system will read the data in the auxiliary data base called "POST" and perform all the adjustments needed to transfer this auxiliary data into the main data base, called "BASE."

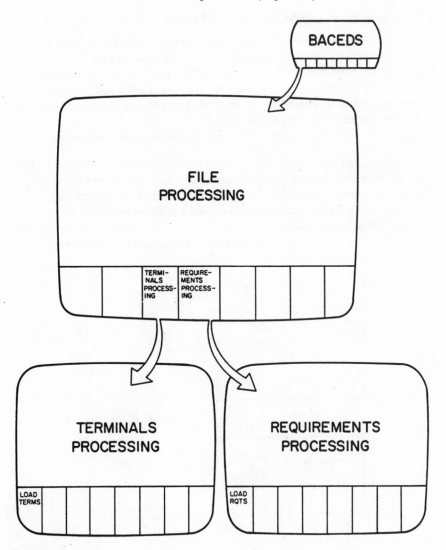

EXHIBIT 8.4.2

8.5   Designing a Module That Teaches the Experienced User

Unlike the tutorial module, which teaches one small item of information to a novice reader, the *demonstration module* presents one whole function, task, or activity. It must be simple and clear—and above all accurate. Experienced readers have little patience with errors and omissions and direct their impatience toward the authors.

As long as they are clearly written and uncluttered, demonstration modules can present substantial chunks of information: complete procedures or transactions, whole programs or modules of programs.

The reader of a such a module expects it to be accurate. That is, if the procedure in the module is imitated, the result should be as promised. If that is not so, the reader blames the writer of the documentation and

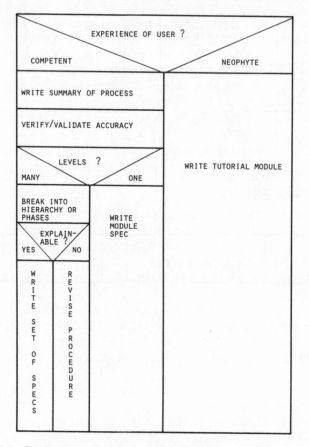

EXHIBIT 8.5   Designing a Demonstration Module

the developer of the system. (This contrasts with the novice users, who tend to blame themselves.)

As Exhibit 8.5 shows, the first task in planning a demonstration module, or a hierarchical series of them, is to be sure the intended reader is an experienced, confident learner, free from the special needs discussed earlier in connection with tutorial modules.

The next task is for the designer of the module to write a summary of the process to be described—usually how the person is supposed to do something. This summary should be terse, a straightforward list of instructions and conditional actions.

Remarkably, many writers of user manuals and operations guides begin to write without actually knowing precisely how to do the thing they are explaining! Not only should the documentors be able to write these summaries, they should then test them for accuracy.

In writing up a procedure that already exists, the test is straightforward: we get someone to follow the instructions (and *only* the instructions) to see if the program or device performs as expected. If the system is still under development, however, the summary of the process must be tested by having the reviewer verify its correctness—a procedure not as good as a live test, but the best possible in the circumstances.

In some special cases, of course, the procedure is part of an early plan or specification; if that is so, there is less a need to verify the procedure than to ensure that the developers create a system that behaves as promised in the spec.

The most interesting document design problem, at this point, is to decide whether the process or transaction is a *one-level* or *multi-level* procedure. In simple terms: to decide whether or not it will fit into one module.

If it *is* one level, if everything that needs to be said about the procedure can be handled in the one-page or two-page module, then the designer writes the summary paragraph, sketches the simple procedural diagram, and considers the module specified.

But what if the whole transaction calls for more than one module, as many do?

If a process is too big for one demonstration module, then it will need a hierarchy of them. That is, it will need one overview module, followed by a series of modules for each main component of the process (a two-level hierarchy); in other cases it may call for a three- or four-level hierarchy.

Defining the hierarchy or components of the process calls for ingenuity; there are always several alternative ways to break a complicated thing into its components. The best way is the one that lets the manual score highest on the *usability index*, that is, the one that reduces the amount of skipping, branching, and looping.

8.5   Designing a Module That Teaches the Experienced User

8.5.1   Example: Demonstration Module for an Administrator

The example below is from the operations manual for a military supplies management system. (It was not written by me.) It shows how to use a particular command, but assumes a great deal of prior orientation. It would be considerably improved with a summary paragaph.

MI

MASTER INVENTORY INQUIRY

PURPOSE:

The MI TRANSACTION is designed for use by:

Inventory Management personnel

Stores personnel at all bases

Maintenance Material Control (MMC) personnel at all bases.

The MI TRANSACTIONS should be used to:

Check for item stock on hand

Check for Supply Organizations (Stores) that stock an item

Check for total inventory information.

DATA BASE USED:

MSDB — Master Supply Data Base

TRANSACTION MODE ALLOWED:

6 — RETRIEVE

TRANSACTION KEY REQUIRED:

Part Number --(32 CHARACTERS)

Manufacturer's Code --(5 CHARACTERS)

or

NIIN --(9 CHARACTERS)

ERROR MESSAGES:

1001 — NIIN ENTERED NOT FOUND
1002 — PART NUMBER/MFG CODE ENTERED NOT FOUND — USE XP TRANSACTION TO CHECK
1003 — PART NUMBER/MFG CODE AND/OR NIIN MUST BE ENTERED
1004 — MULTIPLE NIINS EXIST FOR PART NUMBER/MFG CODE ENTERED — USE XP TRANSACTION TO DETERMINE NIIN DESIRED — REENTER TRANSACTION WITH NIIN
1005 — UNKNOWN PART NUMBER CAN NOT BE ENTERED
1408 — SUPPLY ORGANIZATION INVALID — SEE DATA DICTIONARY
       — NO MORE SUPPLY ORGANIZATIONS TO DISPLAY
       — PRESS ENTER FOR THE NEXT PAGE OF SUPPLY ORGS

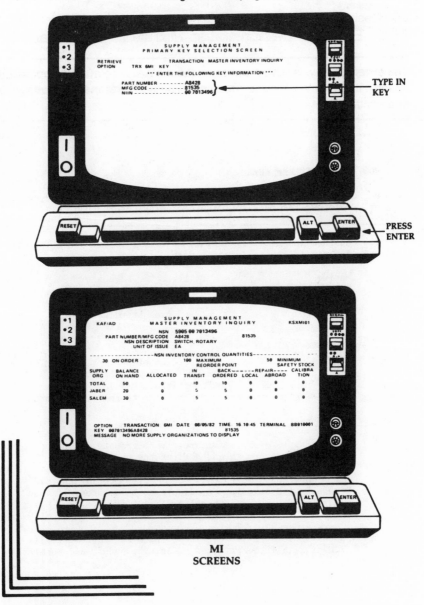

EXHIBIT 8.5.1

8.5 Designing a Module That Teaches the Experienced User
8.5.2 Example: Demonstration Module for a Planner

The excerpt below is from a communications planning manual (not written by me). It shows the planner how to define and redefine the boundaries of a territory.

---

5. Defining Areas

5.1 HOW TO REDEFINE AN AREA

You may redefine existing areas in several ways: one area may become two; or two become one. Any coordinate or side can be changed.

You can redefine areas by adding, deleting, or changing. The system asks for the following information:

Medium	Digitizer or Keyboard
Type of Update	Add, Change, or Delete
Area/Coordinate Update	All coordinates or just some
Area ID	ID of Area being Redefined

Area Updates

Add   System accepts new coordinates until you enter either
       "g" from digitizer or "*" from keyboard
Change Same process; new coordinates overlay old ones
Delete System causes the Master File to remove this area

Coordinate Update

Add   System prompts for present coordinate, then for new one
       The new one is added after the current one.
Change System asks for the coordinates to be changes, then for
       new coordinates. New coordinates overlay the old.
Delete System asks for coordinate to be deleted. If found,
       it is removed and all following it move up.

NOTE: Whenever an area is redefined, the system automatically
      reassigns all the affected buildings, nodes, and terminals.

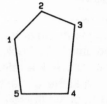

THE ORIGINAL AREA

DEFINED BY POINTS 1 THROUGH 5.

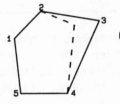

CHANGING ONE POINT

ASSIGNING NEW COORDINATES TO
POINT 3.

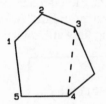

ADDING A POINT

INSERTING A NEW POINT BETWEEN
POINTS 3 AND 4.

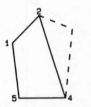

DELETING A POINT

REMOVING POINT 3 FROM THE
DEFINITION OF THE AREA.

FOUR POSSIBILITIES IN CHANGING THE
SHAPE OF AN AREA

EXHIBIT 8.5.2

8.6   Designing Effective Reference Modules

The type of documentation that benefits least from the modular format is reference material: lists, inventories, and compendiums to be "looked up" as needed. The sole criterion for deciding whether to break a reference section into modules is whether this would make the material easier to find and use. If chunking or clustering the material does not aid the reference function, do not do it.

Reference modules give reference—not teaching, motivation, or demonstration. The reference function is to extend the memory of the user: to provide an accessible location for long lists of items that no one ever bothers to memorize, or a convenient access to items that were learned earlier but since forgotten.

As Exhibit 8.6 shows, the first task in designing a reference module or series of them is to assess the suitability of the "standard" presentation, that is, the typical method of presenting long lists and inventories.

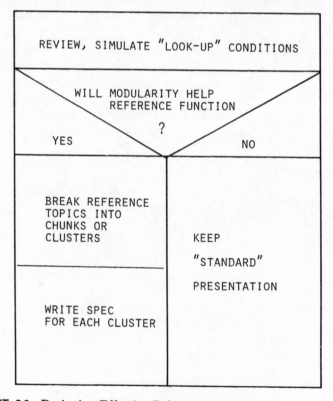

EXHIBIT 8.6   Designing Effective Reference Modules

Should the list be allowed to "wrap around," as the word processing literature puts it, or should it be modularized? For example, is there any advantage in recasting the most familiar reference material – a telephone number directory – into two-page modules?

In many cases there is no advantage. I have seen "logical groupings" of reference lists that worked against the convenience of the reader. For example, one system, with coded error messages divided the reference materials into errors caused by the operator and errors caused by system malfunctions. Unfortunately, though, the operator could not recognize the classification of the error from what appeared on the screen and often had to look in *both* places!

Well-designed reference modules do *not* attempt to teach. One of the earmarks of this module is that it calls for a very short headline, a very short summary, and, often, *no other text besides the summary*. A typical reference module, when finished, will contain nearly two pages of exhibits (charts, tables, lists . . .).

A manual full of reference materials – modular or otherwise – is probably not an effective user manual. Reference is what users need *after* they know how to work the system or product. Until then, reference material is often unfriendly or intimidating.

The most serious violation of this principle is the attempt to *teach in a glossary*. When a manual has been written for user U, U should not have to consult the glossary (which is probably at the front or back) each time a new term is introduced. Glossaries are to help people *remember* what they have been taught in tutorial and demonstration modules. Sending readers to a glossary, or assuming that they will go there frequently, is a way of telling them that this manual was designed for someone else.

Reference modules, alone, cannot teach. Nor should they be embedded inside of teaching materials. It is inconvenient for the user to search a tutorial section in pursuit of a frequently used table.

Rather, reference modules should always be easy to find. They should be at the beginning or end of the manual, even on the covers of the binder. They can be in the form of posters or pull-outs or pages that can be folded pocket-size for "easy reference." Operators often create their own reference materials and keep them in a tiny notebook or even taped to the underside of input devices.

If our users and operators are making their *own* reference documentation – and if we want it to be accurate and maintainable – we had better find out what they need and give it to them.

8.6   Designing Effective Reference Modules
8.6.1   Example: Reference Module for an Executive User

The excerpt below is from an end-user guide to an electronic mail system for a large manufacturing company. (It was not written by me.) Note that it may be used either as an overview or a memory aid, but that it does not in itself teach or demonstrate. Only someone who knows how to use the system will be able to use this chart. (Note, also, that this could have been a one-page module.)

What commands can you use in MAIL?

There are several basic commands used in MAIL--you can read MAIL messages, send messages to other users, delete old messages, etc. Each command is discussed separately in the following pages.

To use MAIL, type in a command--to READ mail, SEND mail, etc.--and press <RETURN>. Commands are always typed at the MAIL> prompt.

EXHIBIT 8.6.1

SUMMARY OF MAIL COMMANDS

Command	Meaning
BACK	Displays the previous message in the directory.
DELETE	Deletes the message most recently displayed.
DIRECTORY	Lists a summary of your messages.
EXIT	Takes you out of MAIL.
FILE	Copies the most recently displayed message into a specified file.
FORWARD	Forwards the most recently displayed message to other users.
HELP	Displays information about how to use MAIL.
NEXT	Skips from the middle of the message currently displayed and displays the next message in the directory.
QUIT	Cancels any deletions you've done and takes you out of MAIL, leaving the directory exactly as it was when you entered MAIL.
READ	Displays the next message in the directory or the next screen of the message you're reading (you can also press <RETURN> to READ).
READ MAIL	Displays a new message recevied while you are in the MAIL system.
REPLY	Sends a message on the same subject to the sender of the message displayed last.
SEARCH	Looks for and displays a message that contains specific text.
SEND	Transmits a message to a user or users.
SEND/LAST	Sends a copy of the message you just sent to another user.

8.6   Designing Effective Reference Modules
8.6.2   Example: Reference Module for a Clerk

The example below is from the Introductory Manual for a commercial software product. (It was not written by me.) It contains a glossary of commands—useful to the word processing clerk or technician who has learned elsewhere how to use this system, or one like it.

10 RUNOFF WORD PROCESSOR

10.2 Command Definition

> RUNOFF processes information stored in a Reality file item. This information consists of both text and RUNOFF commands.

All RUNOFF commands begin with a period (.), and are always on a line by themselves (i.e., the commands do not appear on text lines). Multiple commands may be placed on a line.

RUNOFF processes information in one of two ways. In the "fill" mode, RUNOFF prints a word at a time until no more words can fit on the line. If the justify option is present, RUNOFF adds spaces at random between words to create a right justified margin. This mode is typically used for text in sentence form (the text you are reading now was processed in the "fill and justify" mode). Using this mode, text may be entered free-form without concern for line endings. In the "nofill" mode, RUNOFF processes one line from the item at a time. This is primarily used for tables and figures (such as most of the right hand pages in this manual).

Figure A presents a brief summary of RUNOFF commands. Figure B shows a portion of the item that produced this page.

Command	Description
.BEGIN PAGE (.BP)	Causes a BREAK and a page advance.
.BREAK (.B)	Outputs any partially filled line before processing the next line.
.CAPITALIZE SENTENCES (.CS)	Capitalizes the first word of each sentence.
.CENTER (.C)	Centers the following text line.
.CHAIN ({DICT} file-name item-id)	Chains to the specified text file.
.CHAPTER title	Numbers and formats chapter headings.
.CONTENTS	Prints the table of contents accumulated by preceding CHAPTER and SECTION commands.
.CRT	Specifies output to the user's terminal.
.FILL (.F)	Fills a line without overflowing it.
.FOOTING	Prints the next line on bottom of each page.
.HEADING	Prints the next line at the top of each page.
.INDENT N (.I n)	Indents the next line by 'n' spaces.
.INDENT MARGIN N (.IM n)	Positions the left margin.
.INDEX text	Stores 'text' in an index list.
.INPUT	Reads next line from the terminal.
.JUSTIFY (.J)	Justifies the right margin.
.LEFT MARGIN n	Sets the left margin.
.LINE LENGTH n	Sets the line length.
.LOWER CASE	Causes all letters (except those specified) to be output in lower case.
.LPTR	Directs output to the system printer.
.NOCAPITALIZE SENTENCES (.NCS)	Resets the CAPITALIZE SENTENCES mode.
.NOFILL	Resets the FILL mode.

Figure A. Summary of RUNOFF Commands

.NOJUSTIFY (.NJ)	*Resets the JUSTIFY mode.*
.PAGE NUMBER n	*Sets the current page number.*
.PARAGRAPH n (.P n)	*Formats text into paragraphs (with the first line indented).*
.PRINT INDEX	*Prints the sorted index of words generated by the INDEX command.*
.PRINT	*Displays the next line on the user's terminal.*
.READ ({DICT} file-name item-id)	*Reads and processes the text item indicated.*
.SECTION level title	*Numbers the next section at depth 'n'.*
.SET TABS n{ n}...	*Sets tab positions.*
.SKIP n (.SK n)	*Outputs 'n' blank lines.*
.SPACING n	*Sets the line spacing.*
.STANDARD	*Resets the default parameters.*
.UPPER CASE (.UC)	*Prints characters as they are (lower or upper case).*

Figure A.   Summary of RUNOFF Commands (continued)

```
001 .BP
002 .F.J
003 .READ FOOTING.L
004 .INDEX 'RUNOFF command definition'
005 .SECTION 2 Command Definition
006 .SK 2
007 RUNOFF processes information stored in a Reality file item.
008 This information consists of both text and RUNOFF commands.
009 .SK 2
010 All RUNOFF commands begin with a period (.), and are always
011 on a line by themselves (i.e., the commands do not
012 appear on text lines).
013 Multiple commands may be placed on a line.
014 .SK
015 RUNOFF processes information in one of two ways.
016 In the "fill" mode, RUNOFF prints a word at a time until
017 no more words can fit on the line.
018 If the justify option is present, RUNOFF adds spaces at
019 random between words to create a right justified margin.
020 This mode is typically used for text in sentence form
021 (the text you are reading now was processed in the "fill
022 and justify" mode).
023 Using this mode, text may be entered free-form without concern for
024 line endings.
025 In the "nofill" mode, RUNOFF processes one line from the
026 item at a time.  This is used primarily for tables
027 and figures (such as most of the right hand pages in this manual).
```

Figure B.   Sample RUNOFF Source Item

EXHIBIT 8.6.2

8.7 Mounting the Storyboard

The Module Specs are in a pile—an unworkable form for people who want to test and manipulate them. The next step, then, is to convert the pile of specs to a "gallery" or "storyboard" of specs, by posting them on a wall or two. In this form they can be reviewed and revised by the people who wrote them, the "authors" who will complete them, the users and others who will read the finished manual, and even an official of the organization.

The individual Module Specs are converted to a "gallery" by posting them in the intended sequence on the walls of a room. This process, converting the outline of a book to a visual display, is usually called "storyboarding," a term borrowed from the motion picture industry. (Interestingly, a technique suitable for planning movies is especially suitable for forcing documentors to think of their books as *sequences* of information rather than *hierarchical* collections of information.)

In this form, the one or two people who wrote the specs can really see them for the first time. They "walk through" the gallery, asking each other questions, challenging the emphasis, the scope, and the sequence of the several modules.

Then, the "authors"—all the people who will contribute the missing details to the text and exhibits in the modules—are invited to review the storyboard and make further corrections or suggestions.

Once the planners and authors are satisfied, it is time for the critical user review of the storyboard. Again, actual users, or people who are supposed to know the users, test the storyboard for clarity, accessibility, suitability. They too propose changes.

The designers of the manual should be present when the users or operators (or their representatives) review the storyboard. The questions asked will reveal the clarity and coherence of the design and may also

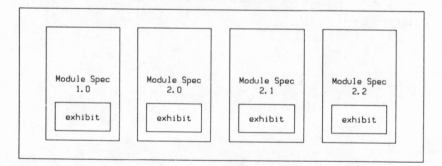

EXHIBIT 8.7   A "Gallery" of Module Specs

correct misimpressions about what the intended readers actually know or do. A storyboard version of a user manual, if prepared early enough in the system development cycle, can actually point out ways to improve the design of the system!

The designers should also *watch* the users and other readers as they review the plan. Unlike the figurative walkthrough associated with structured programming, reviewing the storyboard literally forces people to *walk through* the model of the book. **And the designers of the book can spot design flaws merely by watching the physical movements of the reviewers.** Even at this design stage, many of the nonlinearities of the book – the loops and detours – will be evident as the designers watch the candidate operators and users follow the logic of the manual. The worst design flaws and most unreliable sequences will be evident while there is still time to change them easily.

For the full benefits of storyboarding to be realized, there should be *one* storyboard model, posted in *one* place. In most organizations that is not a problem – aside from the difficulty of finding a spare room in a small facility. But in some larger organizations the various reviewers at interest in the emerging manual are scattered across several sites; in the largest organizations, the documentation or publications unit may be thousands of miles from the developers and users.

Although I am sympathetic to these problems, I still recommend that there be *one* model, in *one* location. As a general principle, I recommend against having more than one review copy of any technical publication, and I also recommend against reviews carried on through the mails. The only thorough technical reviews I have seen were done with all parties present, with lots of questions and discussion, and with all the necessary people and data close at hand.

Eventually, when the designers are satisfied that all the valuable changes have been incorporated, they sign-off the design and invite an official (or official committee) of the organization to review the storyboard. If the design is approved, it is then *frozen*, so that, except for extraordinary cases, no changes will be permitted – unless they are contained within a module. If proposed changes will affect more than one module, we go back to the storyboard.

8.8   Modifying the Storyboard

As in structured design, and as in motion picture planning, one of the most remarkable benefits of this modular storyboard is the ease with which it can be changed. The irony is that the full spectrum of technical and communication flaws that are found in outlines can be addressed with relatively few "design moves." Bad, inaccessible passages—sections that send readers on GOTOs (detours, tangled loops, some ending where they started)—are spotted and corrected merely by manipulation of the Module Specs alone.

The storyboard review, keep in mind, is a *test*, and, like all tests, it must have standards and criteria. A moment ago, I re-emphasized the need to control the number of loops and detours in the manual—GOTOs, in a figurative sense. In some cases, nothing more than such an informal notion or constraint is needed. In an informal review, the goal is to please the reviewers, rather than to meet formal standards.

In a formal review, though, there must be some explicit criteria—especially when there is some dispute about the "best" design or sequence. As a central criterion, I propose the following definition:

> In a *GOTO-less publication or manual*, a reader who begins to read a module will *finish* reading it. If the reader needs or wants more information, he or she will move to the beginning of a new module . . . and finish that one. Moreover, in the most typical case, the second module will immediately follow the first.

In short, this constraint prevents documentors and writers from allowing or compelling readers to leave a module in the middle or enter a module in its middle. It especially prohibits documentors from sending a reader from the middle of one module into the middle of another and back to the exit point in the first one.

Now, as has already been made clear in the earlier discussion of "partitioning" documentation, the more diverse the audience for a particular manual or book, the harder to predict the ways in which its diverse readers will use it. Consequently, it is impossible in principle to develop a book that meets these criteria for all readers. And the greater the diversity of readers, the harder the task.

To the extent possible, however, it is the task of the designers of the manual to recast and rearrange the Module Specs in such a way that the emerging document comes as close as possible to this standard. Every time the book forces a reader to exit or enter a module at the wrong place, we must try to change the design. (Once the first draft is written, it will be too late.)

BEFORE	PROCESS	AFTER
A	DECOMPOSE	A_1 — A_2 — A_3
A — B	CONSOLIDATE	AB
A — B	INSERT	A — X — B
A — B — C	DELETE	A — C
A — B ...→ X	RELOCATE	X — A — B

EXHIBIT 8.8   Modifying the Storyboard

And, surprisingly, even the most complicated changes can be handled with just a few "moves":

DECOMPOSE – converting one module into two or more, in sequence or in hierarchy, with a new spec for each.

CONSOLIDATE – collapsing two or more modules into one, when they are part of the same theme or concept

INSERT – adding one or more modules needed to bridge a gap

DELETE – changing the sequence of two modules, from "logical" to "readable"

RELOCATE – moving a module or group of modules from one place in the book to another

To repeat, these "moves" account for most of the possible changes. (There are also changes *within* modules, which can be effected just by changing the contents of the Module Spec slightly, or by adding notes of emphasis.)

The storyboard technique was invented to ease the process of change and revision. A storyboard plan can be revised radically a dozen times in a day. A full first draft, though, will be patched and plugged, but never really redesigned to eliminate its flaws.

8.9 Won't There Be a Lot of Redundancy?

Ironically, one of the surest signs of success in writing a modular publication is that readers notice—or even complain about—the redundancy. Redundancy across modules reduces the need to branch, loop, or detour. Redundancy within the modules compensates for noise and careless reading.

To sugar-coat the pill somewhat, I could have used some word other than redundancy: something like repetition, or amplification, or reinforcement, or restatement.

But redundancy is what it is: using more than is necessary; spending more than is necessary; writing equivalent information three or four or five times; duplication of sentences, paragraphs, and exhibits. Indeed, given the problem of maintaining user documents, it is better to **repeat identical passages** than to cover the same topics with different words and pictures.

The redundancy in a usable user manual is of two kinds: across modules and within modules. Exhibit 8.9 demonstrates a simple kind of cross-module redundancy. There is a certain procedure that is at the beginning of several other procedures. In a nonredundant publication, the readers would be sent to the initial procedure again and again. (Before learning how to complete Task B, they would be told to read Task A; and the same for Tasks C, D, E, F . . .) Alternatively, in a redundant publication, each later procedure would include an embedded explanation of the startup procedure, repeated identically each time. This practice is familiar, for example, in the manuals for calculators, which usually begin each procedure with a reminder to turn the calculator on and clear its registers.

This issue is complicated and controversial. What is especially interesting is that it suggests a breakdown in the analogy between modular computer programs and modular publications. In certain views, it is the

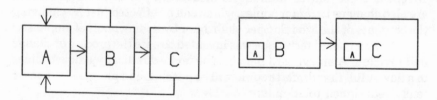

INSTEAD OF: THE REDUNDANT DESIGN IS:

EXHIBIT 8.9   Redundancy Across Modules

essence of a structured program that it is *not* redundant, that whenever a particular task or function occurs it is called from the one place in the program where it resides.

On closer examination, though, the analogy holds up. The real issue is whether the manual we present to the users will have all the "calls" performed for them, or whether we'll expect the reader to search for the appropriate modules and run them at the appropriate times. And the factors affecting the decision are analogous as well. If the recurring material is rather long, it is undesirable to repeat it within each module. If it is going to be invoked or referred to repeatedly, it would add too much complexity and difficulty to the reading process, in much the way that frequent calls add overhead to a computer program.

Redundancy, although it complicates maintenance and seems inefficient and wasteful, reduces the number of skips, jumps, branches, and loops in a publication. For readers with limited book skills, redundancy may be the difference between a usable and an unusable book. And it follows, then, that developers of user manuals may feel safer putting less redundancy into those books intended for sophisticated users who handle complicated publications well.

(Too much redundancy can irritate; excessively repeated directions sound preachy. The amount of repetition in this book, for example, might be more suitable in a user manual than in a text for professional readers. As mentioned earlier, this is meant to illustrate the recommended style.)

Redundancy across modules should use *identical* repetition. Repeated paragraphs and exhibits should be indexed in the word processing system, so that they may be copied exactly and so that when they are changed in one place they will be changed in every place they appear.

In contrast, the redundancy within a module is less often repetition and more often equivalent presentation. Most two-page modules have *three* redundant blocks of information: the headline/summary; the exhibit; and the detailed text – all reinforcing and overlapping each other.

8.10   Handling Branches and Hierarchies

Is it illogical to attempt to design a manual as a sequence of independent modules? What about the branching points and hierarchical relationships? Or what about the long, complicated processes that will not fit into one module?

Sequential communications – in which the ideas and the presentation are at one level, without identation or subordination – are the clearest and most effective. Short training guides, marketing brochures, 20-minute audiovisual briefings, compact proposals – these short, concise communications are best when they are designed with momentum, moving forward urgently as through the steps in an argument or the plateaus in a sale. To the extent that user publications can be made this way, they will be more effective.

But what about the more-complicated technical and reference manuals, hundreds of pages of tables, definitions, and procedures: processes that branch and loop?

There is no problem at all. Through the simple expedient of indenting the headlines, or using an indented numbering scheme, the hierarchy can be not only preserved but **made clearer** to the reader. It is obvious, for example, merely by looking at the heading of the module whether it is above, below, or on the same "level" as the previous module. One can also tell from the information at the beginning of the module what the previous "higher level" was – something that is not apparent in any part of a conventional manual other than the table of contents. (You are now reading a module that is one level below the main heading of Chapter 8.) Modular manuals show their hierarchy and organization in *every module!*

As Exhibit 8.10 shows, the table of contents in a modular publication may be either sequential or hierarchical. The first hierarchy is two-tiered and uses indentation; the second is three-tiered and uses the familiar decimal notation system. Indeed, modular design, by limiting the size of procedures, often *encourages* hierarchy.

Modularity also helps in the presentation of *branching* procedures – activities in which the operator or system veers off on one of two or more alternative paths, depending on the conditions that prevail at the moment. In conventional manuals, these tend to be the most complicated, unreliable moments in the publication. In the modular publication, the best approach is to have the whole procedure, including its alternative paths, contained within one module. Failing that, however, the best approach is to have the module *end at the branching point*. That is, the reader moves to one of the alternative next modules and does not have

EXHIBIT 8.10

Sequential (One-Tier) Outline

Copying the Distribution Disks
Telling the System Your Configuration
Choosing Options and Alternatives
Setting Up a Mailing List
Entering Data into the Mailing List
Revising Data in the Mailing List

Forming a New List with Parts of Other Lists
Printing the Entire List
Selected Parts of the List
Printing Envelopes
Printing Labels
Trouble-Shooting Chart

Hierarchical (Two-Tier) Outline

Four Steps to Get Started
  Copying the Distribution Disks
  Telling the System Your Configuration
  Choosing Options and Alternatives
  Setting Up a Mailing List
Three Ways to Enter an Address
  Entering First Data into the Mailing List
  Revising Old Data in the Mailing List
  Forming a New List with Parts of Other Lists
Printing the List
  Selecting the Addresses to Be Printed
  Printing Envelopes
  Printing Labels
Trouble-Shooting Chart

Hierarchical (Three-Tier) Outline

1. Four Steps to Get Started
  1.1 Copying the Distribution Disks
  1.2 Customizing and Defaulting the System
    1.2.1 Telling the System Your Configuration
    1.2.2 Choosing Options and Alternatives
    1.2.3 Setting Up a Mailing List
2. Several Ways to Create a Mailing List
  2.1 Entering Data from the Keyboard
    2.1.1 Entering First Data into the Mailing List
    2.1.2 Revising Old Data in the Mailing List
  2.2 Using Data from Existing Files
    2.2.1 Forming a New List with Parts of Earlier Lists
    2.2.2 Using Lists from Other Data Bases
3. Printing the List
  3.1 Selecting the Addresses to Be Printed
  3.2 Printing Envelopes
  3.3 Printing Labels
Appendix: Trouble-Shooting Chart

---

to return to the original. Yes, that means some blank pages and yes, that means some redundancy along the separate paths.

Structured design encourages writers to treat their publications as presentations, with the modules in the most usable sequence.

Chapter

9

Assembly: Generating the Draft

9.1   The Advantages of a Frozen, GOTO-Less Design

The main beneficiaries of a GOTO-less publication are the readers. Additionally, though, the people who manage and write publications benefit as well. A GOTO-less design ensures the independence of the modules from one another, allowing them to be written in any sequence, by any arrangement of "authors"; and also allowing them to be reviewed and tested as they come in, without regard for sequence or for the links between the modules.

"Freezing a design," as explained earlier, does *not* mean that the design of the manual will never change. Rather, it means that the design is official and cannot be revised without an official change control routine. No one may make that small change which wrecks the GOTO-less design and, in the process, undermines the independence of the modules.

The GOTO-less manual is a collection of modules in which all the possible connections between modules – all the references and couplings – can be seen in the design. If no reader may be directed to the middle of a module, then the only writer who has to worry about the middle of a particular module is the one who is writing it. Everything that Writer A needs to know about Writer B's module is already in the *storyboard, in the module spec.*

Again, this functional independence among the modules can be lost in an instant if someone departs from, or adds to, the original design without also reworking the storyboard. (Changes that fit entirely within one module and do not affect any of the others are, of course, permitted. Generally, any version of the module that does not call for a new headline or summary paragraph is permitted at the discretion of the writer.)

Further, if the GOTO-less logic is maintained, then the manager of the project can make the best possible use of the writing talent available. People with time to work on only one module (2 to 3 hours of work) can be assigned that single task. People who fall behind can have some of their modules reassigned to others.

An "author" – anyone assigned to write the rest of one or more modules – can write those assigned to him or her in any sequence that is comfortable. A missing item usually cannot delay more than one module; the rest can be written independently.

And, as useful as the GOTO-less design is to the writing of the first draft, it is even more useful in the reviewing and editing of those modules. Put simply, once we know what is in the storyboard, we know enough to review and edit any one of the modules. And if we do not anticipate any changes in the design, we can even assign page numbers and figure numbers to the modules, no matter the order in which they arrive. In-

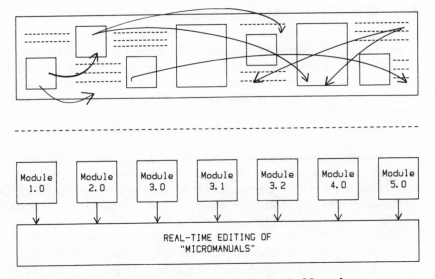

EXHIBIT 9.1   Advantages of a GOTO-Less Design in Manuals

stead of producing a long, tangled series of interwoven paragraphs, the writers produce a series of small self-contained publications, each of which has already been reviewed for its technical content, and each of which fits not only into the logic of the book but into the *physical* form of the book. So, the writers are *implementing*, not *creating*.

One of the effects of developing documentation in this structured style is that it reduces the interest of the first draft. Instead of being the most complicated, demanding, and fascinating part of documentation, writing the draft becomes the least interesting. (Remember, most of the art and intellect has been shifted to the design phase.) Be warned, then, that even though books written by one person still benefit from the structured method, a professional writer will find it boring to carry out his or her own design and may be tempted to wander off onto artistic sidetracks.

The best plan is for two people, a technical expert and a documentor/editor, to design the publication and, then, for them to assign others to carry out the plan.

9.2   Selecting and Managing "Authors"

In this structured approach to user documentation, the first draft is merely the detailing of the storyboard design. In the traditional approach to publication, however, the first draft is the first true attempt to organize and present the content. This difference changes the traditional conception of what "authors" do.

The similarities between programming and documentation are inescapable and instructive. The programmer who works without structured analysis and design goes directly from a vague or intuitive "spec" to the source code. And most documentors – even professional technical writers – make the same mistake.

In the traditional, unstructured approach to manuals and publications, the first draft is a whole piece. At best, it is a few large pieces. Therefore, it is usually considered the large assignment of a single writer or, at best, the collaboration of a very few writers. Because there is so little real specification in the conventional outline, logic demands that the many intricate connections among the parts of the publication be realized by a single person. Lacking external controls, the manual needs the internal control of a single author's mind, assuring that all the parts hang together.

Unless the publication is divided into very large chunks (almost separate manuals), it is nearly impossible to do the work as a team; when the work is divided finely, however, the small parts rarely fit together. The problem is analogous to the problem of "incompatible coding styles" that plagued programming before the ego-less era of structured design.

In contrast, when the publication is fully specified in a set of module specifications, with each module small and independent, and each spec containing all important matters of technical substance, then writing the first draft is an entirely different task. Furthermore, when the sequence of the modules is restricted by a GOTO-less logic, and when the design of the book is frozen, then the writing of the first draft is hardly like what is ordinarily thought of as "writing" at all.

The "first draft" of the emerging manual is produced by having several "authors" supply the missing details in the text and exhibits – *one module at a time*. In principle, there can be as many authors as modules, each working independently. In principle, the first draft of a very long manual could be completed by a team of "authors" within two or three hours of finishing the storyboard!

Even when there are only one or two authors, though, the benefits of the modular design are still impressive. The publication can be prepared in small installments, out of sequence, without worrying about the connections (the "interfaces") across the modules.

A major effect of this structured approach is that it allows for full participation in the writing by even those people who are usually the most reluctant to write. In this scheme, the author is a person who is knowledgeable about the latest details of the module. The author's task is just to provide correct details, within a prescribed space, for material that has already been designed (and reviewed and approved) by every interested party. If a programmer or engineer will not write under these circumstances—especially when told that "English doesn't count"—then he or she will probably not write under any conditions.

Using these unlikely authors not only allows for the rapid completion of the draft, but also improves the technical accuracy of the draft. If the writing is by the most knowledgeable person, the result, though awkward in style, is likely to be accurate. And it also liberates the professional writers to do what they do best: correct, clarify, and improve the writing in the first drafts.

MODULE #	APPRVD BY	DRAFT AUTHOR
1.0	EJCW 10/1	Brown
2.0	EJCW 10/1	Brown
2.1	EJCW 10/1	Brown
2.2	EJCW 10/1	Brown
2.2.1	EJCW 10/5	Lopez *
2.2.2	EJCW 10/5	Blum *
2.3	EJCW 10/1	Gilles
2.3.1	EJCW 10/3	Gilles
2.3.2	EJCW 10/3	Gilles
2.4	EJCW 10/1	Wright *
2.5	EJCW 10/1	Finch
3.0	EJCW 10/1	Jones
3.1	EJCW 10/1	Jones
3.2	EJCW 10/1	Jones
3.3	EJCW 10/5	Calder *
4.0	EJCW 10/1	Finch
4.1	EJCW 10/1	Finch
4.1.1	EJCW 10/5	Finch
4.1.2	EJCW 10/5	Warnier *
4.1.3	EJCW 10/5	Jackson *
4.2	EJCW 10/15	Archive
4.3	EJCW 10/15	Archive
4.4	EJCW 10/15	Archive

EXHIBIT 9.2   Assigning Authors

9.3   Using Project Management to Assemble the First Draft

Although storyboarding and modular design help even the single writer working alone, the benefits escalate rapidly as the size of the project grows or as the number of participants increases. Structured user documentation improves dramatically the *management* of writing, first, by leveling the effort throughout the drafting stage, and, second, by allowing the documentor to use a full range of project management techniques.

Structured user documentation radically alters the nature of assembling a first draft and, in the process, turns documentors into managers.

On any document big enough for two writers, someone must be in charge. But whether the person in charge is a programmer drafted into the job, or even a professional publications manager, there is very little real management in most documentation organizations. Working from conventional outlines, most documentors have little true control over the time and cost of production – and limited control over the quality.

Exhibit 9.3a contrasts the traditional method with the structured method. In the traditional approach, there is a short period of outlining and planning, followed by a long trough of inactivity. During this interval, programmers and writers, who have usually been given writing assignments of undefined scope and size, stall and procrastinate until the deadline – or beyond. Meanwhile, the person in charge hopes ardently to get back some drafts, usually in vain.

Naturally, the person-in-charge ends up doing much of the writing alone, usually on a crash schedule, with little opportunity to edit and revise.

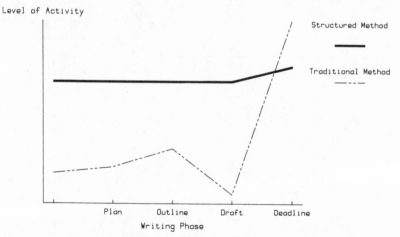

EXHIBIT 9.3a   Contrasting Dynamics of Traditional and Structured Publications

In contrast, a structured manual takes longer to plan and design. But within only a few hours of finishing the outline, the first draft versions of the modules (all of which are independent and small) must come back to the one in charge. There is a *level effort* throughout the process, time to edit, test, and revise.

If a particular author fails to respond in the time needed to write a module, the documentor can investigate at once — perhaps with the result that the module is assigned to another author, while there is still plenty of time.

This shifting of the paths through the production network demonstrates another advantage of structured documentation: each module is a well-defined parcel of work and can be placed in a project network. Each module is a task of defined size, with a person in charge, an estimated duration, and, in some cases, a budget for artwork and production. The manager can estimate the costs beforehand and, by manipulating the assignment of modules, can predict and adjust the completion date. In the best case, all the modules are independent, so that the only constraint in the network is the result of having one author write more than one module.

The more thoroughly enforced the modular design, the greater the opportunities to honor budgets and shorten the "critical path" of the production plan. And the more the people who supervise writers (amateur or professional) can exploit some of the new project management tools currently available for small computers.

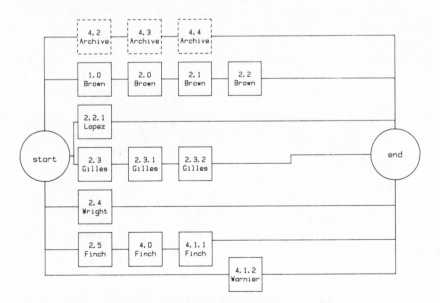

EXHIBIT 9.3b   Publication Production Network

Chapter

10

Editing: Testing and Revising for Readability and Clarity

10.1   Testing the Draft: Main Issues

The purpose of testing the draft is to "edit out" the tactical errors. The goal is to correct as many as possible of the awkward and incorrect passages that cause readers either to make mistakes or to reread the difficult or ambiguous sentences.

An "unstructured" first draft, written from a traditional outline or specification, is usually so difficult to read, and so filled with new technical material, that it is in urgent need of technical review. If time permits, there will be a style review and another quick technical review. A structured manual, however, can be reviewed module-by-module. Because the technical content of the module has already been reviewed, it is possible to clean up the language and presentation first, and then later do a light technical review to catch matters of detail.

The traditional first draft of a manual is about 90% new — material that no responsible person has reviewed or tested before. Under these conditions, the logical thing to do is to begin with a meticulous review of the technical accuracy of the draft — a process made difficult by the unedited, first-draft prose. There is virtually no opportunity to make the language or artwork clearer, more readable, or more effective.

In contrast, each module of a structured manual is a self-contained

EXHIBIT 10.1   Alternative Ways to Test a Draft

If the draft is monolithic . . .	If the draft is modular . . .
Then, for the Whole First Draft	Then, for Each Module
Initial Technical Review • first reading of complicated text • discovery of major errors • attempt to separate technical errors from confused, ambiguous language	*Initial Language Review* • revision of raw draft language • clarification of ambiguous content • identification of gaps and inconsistencies
Quick Language Review • hasty review of mechanics • minor editorial cleanup • writing specs for printing/ production	*First Technical Review* • verification of technical content NOT already tested in storyboard • incoporation of late technical changes
Final Technical Review • rushed, perfunctory double-check • clumsy updating (errata pages)	*Final Language Review* • refining language/art for usability • careful production and proofing

micro-manual. The technical content and logical connections have been specified by a technical expert and reviewed by other experts. Because there are no big technical surprises in the module, and because the review is not delayed until the last moment, it is wise to **begin with a language and art review. It is easier to find technical errors in a clearly written (that is, edited) text than in a first draft.**

The purpose of this language review is to assure that the document is *clear*, free from ambiguity and misleading information, and *readable*, no harder to read than it must be, and certainly not too hard for the intended reader. In general, the more resources spent on editing a draft, the clearer and more readable it becomes.

All first drafts, even those written by professional writers, need editing to become usable. The editorial improvements are usually in five categories:

- *Mechanics*, correction of errors of grammar, spelling, and punctuation—essential for any firm or individual that wants to avoid embarrassment
- *Appropriateness of language*, replacement of words and phrases that are unfamiliar to the reader, unnecessarily difficult, wrong in tone
- *Clarity*, replacement of words, phrases, constructions, or graphics that have several possible meanings, or that mislead the reader
- *Accessibility*, elimination of awkward, wordy constructions and difficult artwork that do little more than "show off" and, therefore, inconvenience the reader; elimination of the first-draft commonplaces, such as backward sentences
- *Urgency*, revision to make the writing more interesting and engaging, through careful diction, close editing, variations in sentence length and style, and graphic skill

In case someone in the organization complains about the time spent on editing, explain that unclear sentences mask technical errors and invite trouble. For example, the sentence

Managers are required to sign off on Form A51 to approve continuation of a project.

is a horrible mess. And the worst thing about it is that it seems to require or oblige managers to sign a particular form—which is not intended. Applying principles explained later, you may convert the sentence to

If managers want to continue the project, they must sign Form A51.

10.2   Editing for Word and Phrase Bugs

The easiest improvement to make is removing or revising certain common word and phrase bugs, notably: showing off with long or fashionable words; using too many words; using too few words; and putting words and phrases into the wrong places.

Certain commonplace errors of style have the double effect of, first, masking the technical errors and omissions in the text, and, second, increasing the chances that the reader will reread or misread your explanations.

Showing off consists in using long words where short, familiar words would have been just as effective. For instance, *utilize* for *use*; *facilitate* for *help*; *initiation* for *start*; even *depress* for *press*.

Do not misunderstand. There is no reason to write in one-syllable words, and there is no advantage in replacing a technically correct word with a shorter, incorrect word. But there is no gain in using *indicate* for *show*, or *disseminate* for *send*, or *effectuate* for *cause*.

Another common form of showing off is the use of vogue words (buzz-words), like *capability* for *ability*, or *prioritize* for *rank*. The word *environment* is used so often and so carelessly that I have seen it used with opposite meanings in the same publication. Also beware of *transparent* which means "invisible" to computer people and "obvious" to speakers of business English.

Using too many words is a technique youngsters learn as a way of stretching a 300-word idea into a 1000-word composition. A few examples will demonstrate:

- should it prove to be the case that = if
- by means of the utilization of = with
- at that earlier point in time = then
- conduct an inspection = inspect
- perform the calculation of the projections = project

Wordiness is inevitable in first drafts. The most frequent offenders are the "smothered verb" (*make a distinction* for *distinguish, accomplish linkage between* for *link*); the phrase used where a single word would do (*in order to* for *to, with regard to the subject of* for *about*); and, occasionally, clauses for phrases:

BEFORE: After we had approved the test plan, we began the . . .
AFTER: Having approved the test plan, we began the . . .

Using too few words is found often in the writing of engineers and computer programmers. Driven by a desire to be concise, some writers produce phrases and sentences that are compressed to the point of incomprehensibility. What does it mean, for example, to say that a certain system has an "English-like report generating capability"? What are "contiguous sector reference designators"?

No one but the author is sure what is meant by "operational planning materials format design criteria" or "management responsibility assignment history file." Most people cannot understand strings of nouns, or what the programmer would call "noun strings." And adding a few modifiers does not help.

People who avoid prepositions and who cram words together in this cryptic way also tend to eliminate other "useless" words like *the* and *a*. If a *that* is optional they remove it. Indeed, they tend to leave out all the optional commas as well. The trouble with these zealous choppers and cutters is that they destroy the flow of the English language and produce sentences that, though they may be shorter, also take longer to read.

Misplacing words and phrases can also throw your readers off the track. In English, modifiers should be next to the words they modify — usually before. But, in most first drafts, many of the modifiers tend to be misplaced, notably *only, nearly, almost, already, even,* and *just.* In the instruction

Only enter the hourly rate for exempt employees.

there are three alternative ways to interpret the *only.* (Only the *hourly* rate; the hourly rate and *nothing else*; or the hourly rate only for *exempt* employees). Be careful of these modifiers. Do not write

The system *nearly prints* everyone's checks.

when you mean

The system prints *nearly everyone's* checks.

Similarly, descriptive phrases should be near the word or phrase they describe. What does this instruction mean?

Report every unauthorized access *in keeping with company policy?*

Does this mean that if the unauthorized access is not in keeping with company policy that you should not report it?

10.3   Editing for Sentence Bugs

Although there are dozens of things that can go wrong with a sentence, the five flaws most likely to stop or distract a reader are: backwards construction; meaningless predicates; tangled passives; dangling introductory phrases; and marathons.

The secret of the readable sentence is that the "payload" of the sentence – the material that the author would have underscored (if something had to be underscored) – is **at the end.** And what could be more logical? If a sentence is long, it is read and processed in stages; the last read part is the best remembered part. If something critical was at the beginning, the reader would have to "loop" on the sentence.

Even an amateur writer can learn the drill: review the first draft of the sentence; see if the material to be emphasized is at the end; if not, rework the sentence to move it to the end – unless there is some technical reason for not doing so. So, we convert

Reduced cost is the main advantage of this new procedure.

to

The main advantage of this new procedure is reduced cost.

Similarly, when we edit instructions we put the key material last. And if it is a conditional instruction (*if-then*), we make sure that the *then* clause is last.

NO: DFIL is typed.
YES: Type DFIL.

NO: Type DFIL to see what file names have been assigned.
YES: To see what file names have been assigned, type DFIL.

If the payload of a sentence is at the end, then it follows that the "action" in the sentence is usually in the *predicate*, not the subject. Yet, not only do many writers put their main material at the beginning, they sometimes say everything interesting before they even get to the verb, leaving a *meaningless predicate* (what some editors call the "vitiated" predicate).

Consider: "The possibility of underpricing by the Japanese exists." The entire predicate of the sentence is the word *exists*. But to revise the sentence we have to know what the writer wants us to understand. Is it (1) "The Japanese may underprice us." or (2) "We may be underpriced by the Japanese."? Both sentences are grammatically correct. Sentence 1 emphasizes *us*; sentence 2 emphasizes *Japanese*.

English is filled with devices that allow the editor to move phrases from front to back. Among the most useful is the passive form of the verb. Converting (1) "ZAKO Industries acquired an XTRON." to (2) "An XTRON was acquired by ZAKO Industries." changes the verb from active to passive and changes the emphasized word!

Most editors and teachers of writing, though, warn amateurs against the passive form—and with just cause. Tangled passives can ruin an otherwise understandable passage. Consider these pairs:

PASSIVE: Insufficient flexibility is exhibited by the system.
ACTIVE: The system is too inflexible.

PASSIVE: Cheap collating and binding are accomplished by this device.
ACTIVE: This device collates and binds cheaply.

Passive constructions are typically wordy and difficult. But they can, when used carefully, help you to propel the "payload" of a sentence to the most effective position: the end.

Still another way to push the main stuff to the end is to use introductory phrases. (Nearly all conditional instructions have introductory phrases.) The danger in these is the dangling introductory phrase, a string of words disconnected from the body of the sentence.

Again, the drill is simple. The introductory phrase must be tied to the grammatical subject, which should appear right after the comma.

NO: With your simple payroll requirements, PAAY is the system for you.
YES: With your simple payroll requirements, *you* should use the PAAY system.

NO: To locate definitions quickly, glossaries are posted at each work station.
YES: To locate definitions quickly, *operators* can use the glossaries posted at each work station.

NO: When coldstarting the system, the operating system tape is loaded.
YES: When coldstarting the system, *(you)* load the operating system tape.

There are also some strange danglers at the ends of sentences. Beware of such absurdities as: "Do not service the printers while smoking."

Finally, someone must be sure that the sentences simply do not run on too long. The problem is with *marathon sentences* in general, especially several in a row. No one can handle the sentence below:

In addition to solid, dashed, phantom, centerline, and invisible line fonts, numerous linestring fonts are available that provide generation about a centerline with variable spacing (width), layer of insertion options, and left, right, and center justifications.

10.4   Ten Ways to Write an Unclear Instruction

Any word, phrase, or sentence bug can hurt clarity and usability. And the consequences of unclear instructions can be expensive.

1. *Long, vogue words.*

   BEFORE: In the Information Center environment, the manager should utilize a prioritization ranking to facilitate equitable scheduling.

   AFTER: In the Information Center, the manager ranks each job to yield a fair schedule.

2. *Too many words.*

   BEFORE: In the event that you have a lack of knowledge regarding which files you have permission to write in, make use of the PRIFIL command.

   AFTER: If you do not know which files you may write in, use the PRIFIL command.

3. *Too few words.*

   BEFORE: Column heading revision permission may be obtained by HCOL entry.

   AFTER: To get permission to change the headings of the columns, enter HCOL.

4. *Misplaced words/phrases.*

   BEFORE: Only write corrections, not changes, on the worksheet.

   AFTER: On the worksheet, write only corrections, not changes.

5. *Backwards construction.*

   BEFORE: Press the *Clear Rest* key if you want to erase everything after the cursor.

   AFTER: If you want to erase everything after the cursor, press the *Clear Rest* key.

6. *Meaningless predicate.*

   BEFORE: The efficiency of spot-checking the coding sheets before commencing keypunching is worthy of mention.

   AFTER: It is efficient to spot-check the coding sheets before you start to keypunch.

7. *Tangled passive.*

   BEFORE: Care must be exercised in sending sensitive data.

   AFTER: Send sensitive data carefully.

8. *Danglers.*

   BEFORE: When reconciling the account, the encumbrance file must be frozen.

   AFTER: When reconciling the account, (you must) freeze the encumbrance file.

9. *The unnecessary third person.*

   BEFORE: The operator then enters his or her security status.

   AFTER: Enter your security status.

10. *"Cowardly recommendation."*

BEFORE: It is a requirement that operators receive 40 hours of instruction before they enter any real data.

AFTER: Operators must receive 40 hours of instruction . . .

EXHIBIT 10.4a   Original Instructions

1. If your configuration has sufficient RAM capacity, you may utilize the system's windowing capability.
2. Should it prove to be the case that you have some reservations regarding the forecasts, you have the option of using alternative discount rates.
3. Early manual design yields procedural usability benefits.
4. The slide-maker only can be used by systems with 512K memory and hard disks.
5. Type PINSTALL to change the printing options.
6. The urgent need to save data at least every ten minutes is called to your attention.
7. File linkage can be accomplished by key specification.
8. To call the Calculator, <alt> and <c> must be pressed.
9. The clerk should then type the number of the desired file.
10. It is the responsibility of the arriving operator to read the trouble report from the latest shift.

EXHIBIT 10.4b   Corrected Instructions

1. If your computer has enough memory you can use the *window* feature.
2. If you doubt the forecasts, you may try other discount rates.
3. Writing manuals early makes the procedures easier to use.
4. The slide-maker can be used only by systems with 512K memory OR hard disks.
5. To change the printing options, type PINSTALL.
6. You must save the data at least every ten minutes.
7. To link the files, specify the keys.
8. To call the Calculator, (you) press <alt> and <c>.
9. Type the number of the file you want.
10. The arriving operator must read the trouble report from the latest shift.

10.5   Editing to Make the Book Easier to Read

The term *readability* has been coined to refer to the difficulty of a particular text. The word "difficulty" here refers to the sheer effort needed to read a passage. The most popular of the many indexes of readability are Robert Gunning's Fog Index, a simple technique for approximating the "grade level of difficulty" of a passage, and Rudolph Flesch's READ Scale.

Today, there are dozens of candidate indexes or scales that purport to measure readability and, at last count, Bell Labs' "Writer's Workbench" included five different scales.

All readability scales are rough; probably, any one can be faked. Everyone has seen ingeniously composed passages that scored "easy" on the readability scales but were, obviously, nearly impossible to read. The purpose of these metrics is to extract some "objective" assessment of how hard a passage is for the reader to process. The most popular scales usually contrive to have the score equal the "grade level" of difficulty, that is, the years of schooling needed to read the passage.

The best known, Robert Gunning's Fog Index, adds the average number of words in a sentence to the percentage of "hard" words and multiplies by a constant (.4) to yield the Fog Index. (In Gunning's scheme, a "hard word" is any word with three or more syllables, except for proper names, compounds of simple words, and three-syllable words in which the third syllable is *ed* or *es*.)

To test the Index, consider this passage, published by one of the world's largest manufacturers of hardware and software:

> Today's advancements in educational management combined with the rapid growth in student enrollment in schools has emphasized the need for data processors to be used in establishing and maintaining a student records data base, required for providing attendance and academic mark reporting data to satisfy several disciplines. The purpose of this program product is to provide a systematic procedure for recording, retrieving, manipulating, and reporting significant student data, such as attendance and academic mark information. One of the objectives of this program is to provide effective data on individual students as well as aggregate, statistical reports needed for sound analytical decisions by educators and administrators.

This passage has 105 words, 3 sentences, and 34 "hard" words, according to the Gunning criteria. Its Fog Index is .4(35 + 32) = .4(67) = 26.8

There is no person on earth who can read this passage without difficulty! And this is especially unfortunate when you realize that the passage says very little indeed, and could easily have been revised to the 10 or 11 level.

(Another popular measure is used by, among others, the U.S. military in testing the reading difficulty of its manuals. This favored military scale is a revision of the Rudolph Flesch READ scale, altered by the military so that it, too, reports grade level. According to the Revised Flesch Scale, the passage above is at the 21.6 level.)

Of course, merely lowering the readability score of a passage does not solve all its problems. A text with a score of 6 or 7 can still be unintelligible. Whatever quarrel one might have with these particular indexes – or even with the entire concept of simple readability measurement – there is no denying that excessive difficulty ensures that most readers will be unable to make sense of manuals. Even if the manuals are clear, correct, and well designed to eliminate GOTOs, they may still prove unreadable.

No technical publication need score higher than 14 on the Fog Index. No general business publication need score higher than 12. And no manual aimed at clerks and junior technicians should score higher than 8. The simpler a technical publication, the more people there are who can read it.

EXHIBIT 10.5   Two Readability Measures

The Fog Index (Gunning):
Grade Level of Difficulty = .4(average words per sentence +
percentage of hard* words)

*Hard Words = all words with three or more syllables, except

- proper names
- compounds of small words
- three-syllable words in which the third syllable is *ed* or *es*, which would otherwise have had only two syllables

---

The Flesch READ Scale (Adjusted by the U.S. Military):
Grade Level of Difficulty = [.39(average words/sentence) +
11.8(average syllables/word)] − 15.59

10.6   Demonstration: Procedures Before and After

To illustrate the effects of editing the draft, two actual passages from user documents are shown "before" and "after." Several editorial improvements are made in each, most notably a dramatic reduction in the reading difficulty, as measured on the Fog Index.

The passage below comes from a real manual, the project development guidelines for a large international financial institution.

BEFORE:

Following identification of needs and appropriate preliminary approval of all major system development project proposals, the Information Systems Department will prepare an analysis and recommendation for action. The more routine requests will be approved by concurrence of the Information Systems Department and of the financial area management without further review. Those requiring a change in policy, exceeding the approved budgets or crossing organizational lines will require review and approval by the Steering Committee as well.

The Information Systems Department will evaluate the capability of the user or regional technical staff to implement a proposed system. Based on this evaluation, the responsibilities and authorities of the Information Systems Department, regional technical staff, and the user will be outlined in a system development proposal submitted to the Steering Committee.

Words: 127   Sentences: 5   "Hard" words: 37   Fog Index: 21.6

With a bit of editing, an exceedingly difficult (though typical) bit of administrative procedure becomes easy enough for any business professional to follow.

AFTER:

First, needs are identified and major development proposals get preliminary approval. Then, the Information Systems Department analyzes each request and recommends an action.

For small, routine requests the Information Systems Department and the manager of the functional area may approve the project without further review. (A project is "routine" if it does not call for a change in policy, exceed current budgets, or cross organization lines.)

For major requests, though, the Steering Committee must also approve. To advise them, the Information System Department submits its own evaluation, which proposes schedules and tasks for all the participants.

Words: 97   Sentences: 6   "Hard" words: 12   Fog Index: 11.4

In the next case, the "before" comes from the FORTRAN programmer's guide published by a large time-sharing company.

BEFORE:
It is critical that variables used as subscripts in FORTRAN programs always be consistent with information declared in the DIMENSION statements. Unless checking is specifically requested, subscript ranges are not checked for validity when programs are run. This checking is omitted in order to maximize running-time efficiency. However, if invalid values are used for subscript variables, such as a value less than one or greater than the maximum subscript as specified in the DIMENSION statement, errors can occur. Often such errors either go undetected or cause apparently unrelated failures and diagnostics.

When invoking the FORTRAN compiler, the user can inform the compiler that subscripts are to be checked for range validity by supplying the SUBCHK option.

Words: 115   Sentences: 6   "Hard" words: 21   Fog Index: 14.8

In the "before" form, this passage is understandable to a good reader after two or three attempts. The programmers who get it on one reading are those who already know what it is about. The "after" version, however, without "talking down" to anyone, brings the material into the range of most of the English-speaking adults who might choose to read this passage.

AFTER:
Variables used as subscripts in FORTRAN programs *must* stay within the range of those in the DIMENSION statements. (That is, the value of the variable must not be less than 1 or greater than the highest subscript in the DIMENSION statement.) If they are out of the range, invalid, the mistake is often overlooked. Worse, these errors often cause "unrelated" failures or odd diagnostic messages.

To save running time, this system does not check the range of the variables unless told to. To be safe, when you invoke the FORTRAN compiler, tell it to validate the values with the SUBCHK option.

Words: 100   Sentences: 6   "Hard" words: 8   Fog Index: 9.8

10.7   Other Ways to Make a Manual More Accessible

To make a manual accessible, the documentors should eliminate as many distractions as possible, present the material in a package that communicates confidence, and lay out the pages effectively.

Each publication or other information product should be *freed from distractions.* Most distractions are mechanical and production errors: mistakes in spelling, punctuation, or grammar; inconsistent conventions and terminology; acronyms and abbreviations that are not spelled out often enough; awkward layouts; poor photography or color separation. Each occurrence of these bugs, though not likely to do much harm in itself, can cause just enough distraction and confusion to undermine an instruction or break a reader's concentration. Moreover, a persistent, chronic pattern of such bugs can imply an attitude of carelessness or sloppiness. And that is simply the worst possible message to send to readers of manuals.

Along these lines, it is also important to do everything possible to *communicate confidence* to the reader. Careful editing for small bugs will help. So will high-quality typing, copying, and binding. Expensive paper may be the hardest aspect of documentation to justify, but it does, unquestionably, create a better response in users and customers than does cheap paper.

If documents are produced by a printer – the machine, not the person – be sure that the paper is heavy enough so that the typed characters do not "bleed through" the back. If your copy machine is a "bargain," be sure your pages do not look like a "bargain."

Be warned that anything that looks cheap or chintzy may undermine the effectiveness of a document. Usually, it is just a matter of taking away the reader's respect: the user does not take seriously what the documentor did not take seriously. Often, though, the cheapness produces material that is nearly inaccessible. For example, the practice of cramming as many words as possible onto a page – the refusal to use large type, bold face, "lettering," or any other form of more-expensive typography – produces manuals that are torture.

Further, documentors must be wary of any manager whose principal objective seems to be to conserve paper. **There is no communication benefit in saving paper.** Wide margins and big print are better for readers – all readers. An uncluttered page is a page less likely to produce fatigue, and, therefore, less likely to encourage errors. Thick paper, good binders, tabs between the sections, better typography, color – none is essential, but all can help a system realize its full usability.

Documentors must also be wary of the brand of editor whose objective seems to be to save paper by the reckless elimination of words and the incessant use of abbreviations and other compressed forms. There is a profound distinction between clear, concise writing, on the one hand, and compressed, impenetrable writing on the other. (An editor who would cut *on the one hand . . . on the other* from the last sentence does not understand the point.)

Ultimately, firms and organizations that produce lots of publications must acquire competent, professional editors. Programmers can be taught to write a little better; word processing systems can be taught to catch mechanical errors and compute a Fog Index. But there is still a need for someone who knows that good communication demands patience and repetition, someone who knows the difference between conciseness and denseness, between compactness and clutter. And, perhaps, someone who knows that the word "prioritization" is just plain ugly.

EXHIBIT 10.7   Saving Paper versus Increasing Readability

Conserves Paper	Helps the Reader
Narrow margins	Wide margins
Small type, dense layout	Larger, varied type sizes
Minimal illustrations and exhibits	Frequent, large charts, art
Run-on, wraparound printing	New page for each new section
Austerity, minimal explanation	Redundancy, accessibility tools
Typewriter headings/graphics	Typeset headings, professional graphics

11

Maintenance: Supporting and Updating User Manuals

11.1   Maintaining Documents: Stimulus and Response

All documents will need changes. In fact, it is a truism that every document your organization has published, including those published quite recently, needs changes. The most important consideration at this stage is to regulate and control the distribution of documents and supplements. Think of each impulse to change or revise a document as a stimulus; and think of the rule governing the correct response as the maintenance standard or policy.

As with programs and systems, the first requirement in assuring that a manual or set of documents is maintained is to **assign someone to the task.** Unless some person takes the responsibility, the job probably will not be done. For every document, someone must be assigned the task of seeing that it is distributed correctly and that updates and supplements are sent to the right people at the right time.

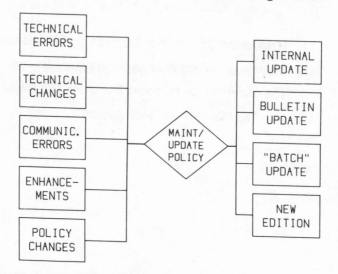

EXHIBIT 11.1 Stimulus and Response

All manuals will need to be changed. As Exhibit 11.1 shows, no matter how carefully they are reviewed, your manuals must respond to such stimuli as

- *technical errors* – incorrect or incomplete technical information about the system or product
- *technical changes* – minor modifications in the system, made while you were preparing the documentation, with or without your knowledge

- *communication errors* – ambiguous, unclear, or misleading text and diagrams in your manual; errors of grammar or mechanics
- *system enhancements* – major changes and new features added to the product or system, scheduled or "ad hoc"
- *policy changes* – new rules on what must or may be done, by whom

To repeat, someone must feel responsible for keeping track of these problems and doing what needs to be done about them. But, interestingly, that responsibility need not result in an endless stream of up-to-the-minute bulletins, warnings, and releases.

The most important thing to remember is that, despite the documentor's understandable wish to have all manuals current and correct, **all manuals are out of date anyway.** The question is not: How can the manuals be instantly updated or corrected? The question is: Which changes can be held for a while – "batched" – and which must be communicated at once? Of those that are batched, which can be held for only a few days or weeks? Which for several months?

In fact, there are four main ways to respond to a stimulus:

- *Internal change* is a correction in the master version of the document, that is, the material kept in the files of the person responsible for maintenance; this file contains modifications in the documents, areas that need modification, and release schedules. Everything in this internal file is urgent and should be kept as current as possible.
- *Immediate update* is sending a hot bulletin to every user or manual owner; obviously, it is a tactic that should be reserved for important messages. Why? Because a flurry of emergency bulletins creates confusion and gives the impression either that your system is in chaos, or that you "cry wolf."
- *"Batch" update* is the collection of several changes in one set, published by the calendar (once-per-month or once-per-quarter) or when the quantity of the material exceeds a certain threshold or criterion.
- *New edition* is the ultimate batch update; again, either according to the calendar or the sheer bulk of the changes, the documentor incorporates all the modifications since the last edition into a new edition, removing what is obsolete, replacing what has been modified. Then the users or DP centers get a new edition (ideally, only after they have handed back the old one).

There are, then, several ways to respond, several different levels of urgency (as in all engineering problems). Still, though, many zealous documentors are too eager and, often, create more confusion than clarity with their incessant updates.

11.2   Improving Maintenance by Anticipatory design

For manuals and documents to be maintainable and modifiable, they must have been designed that way. Modular publications, tested when they are still models, not only reduce the need for subsequent changes, they simplify the process of making and controlling whatever changes must be made.

It is an axiom of software engineering that programs are made maintainable in their early stages of design; after a program has been coded, it can be maintained, but not made more maintainable. Attempts to "retrofit" the system can be ingenious, or even heroic. But they are scarcely as effective as having done it right the first time!

Maintainability is an aggregate measure of the ease and speed with which bugs, flaws, and other inadequacies in a system can be located, defined, and corrected. The maintainability of a system, in most cases, is the single most important predictor of its cost-effectiveness.

Similarly, the manuals and other information products that accompany computer systems also must be maintained and modified. The systems change or develop bugs, and the accompanying user documentation must be changed. Or, the documents themselves can manifest bugs and weaknesses that are independent of the systems they accompany. In time, we shall realize that maintaining and modifying manuals cost more than writing them – and that manuals which resist maintenance are likely to fail, substantially reducing the usability of the systems they support.

The most straightforward way to maintain a manual is to, first, find the modules that need to be changed and, second, repair or replace them. For modifications, this may also entail finding the right place to add a module and, then, inserting it.

When user documentation is modular and structured, one can maintain a directory of all the modules, coded so as to define the systems, topics, applications, installations, or other descriptors that are relevant. Clearly, such a scheme will allow us to search the file for all the modules affected by a particular system change. And it will also allow us to generate new documents from old modules.

If manuals are viewed as unique sets of modules, then one can maintain a directory like that illustrated in Exhibit 11.2a. Because it is likely that a module will appear in more than one place, such a directory tells us *all* the publications that are affected by a particular change in the system.

In large and sophisticated organizations there may be alternative versions of the same technical content expressed in "equivalent modules"; the directory illustrated in Exhibit 11.2b allows the documentor to map

EXHIBIT 11.2a   Directory Record for a Module of Documentation

Module Name: <u>Adding a Record</u>   Module File No.: <u>B-008</u>

First 60 Characters:
   To add a record, type the name of the file in the GOTO win
Superordinate Modules: <u>B-002 Using the Four File Transactions</u>
Subordinate Modules: <u>B-028 Trying to Add a Record That Already Exists</u>
              <u>B-029 Trying to Add a Record With Key Data Missing</u>

Descriptors:
   Program/System: DB-3, Real-Estate Manager, Loan-Manager
   User Tasks: file creation, file update, new account, new record
   Audience: end user, realtor, loan officer
   Site/Installation: ABCO Finance, Goldschmidt & Wong Real Estate
   Other:
Publications/Products in Which the Module Appears:
   G-3, G-4, G-5; F-1, F-2, F-7; R-1, R-5

EXHIBIT 11.2b   Directory of Published Modules

Module	Equivalent Modules	Publications/Products
B-008	R-006, D-120	G-3, G-4, G-5, F-1, F-2, F-7, R-1, R-5
B-028	R-061, D-121	G-3, G-4, G-5, F-7, R-5
B-029	R-062, D-122	G-3, G-4, G-5, F-7
C-110		G-2, G-3, G-4, G-5
C-115	R-090	G-4, G-5, R-5
C-240	D-600	G-3, G-4, G-5, F-7, R-4, R-5
•	•	•
•	•	•
•	•	•

all the consequences of a technical change onto the various publications that need to be changed.

These illustrations are, of course, rather complicated and ambitious. There are other, simpler anticipatory design choices that can make your documents more maintainable. For example, manuals in loose-leaf binders, obviously, are more susceptible to change than bound books. (On the other hand, though, you can prevent certain kinds of misinformation by deliberately *not* using loose-leaf manuals.)

One-page modules, printed on one side of the paper, are the easiest to add, remove, and insert. And they are probably the best form of module for documents that need to be changed continuously. On the other hand, though, they tend to make many manuals choppy and filled with complicated references and loops. The maintenance advantages of the one-page module—with text and exhibits on the same page—may have to be traded-off for the reliability and readability advantages of the two-page module.

11.3   The Maintenance Paradox: The More the Messier

Absent a thoughtful policy for the maintenance of documents, there tends to be a random or haphazard distribution of supplements, bulletins, releases, and updates. What many do not realize is that each supplement to a manual can actually double the number of alternative versions in circulation. And only one of these versions is correct.

Again, it is a truism that all manuals are out-of-date and that they contain at least a few errors. This is no more remarkable or deniable than the claim that all complicated programs or devices have bugs – including some that have not yet been recognized.

It is also a truism that all user lists, distribution lists, and route lists contain errors. And the longer the list, the more inaccurate and out of date. That is, any attempt to communicate bulletins and changes to all the people who are using a certain document – or a certain system – will be frustrated by the inaccuracy of that list. The exception, of course, is when the programmer, operator, and user are all the same person. In those circumstances, though, there is rarely any formal documentation anyway.

These first two truisms – that all documents are inaccurate and that all distribution lists are inaccurate – are almost natural laws of technical communication. If we add another law, the Second Law of Thermodynamics (the entropy principle), it becomes more understandable why attempts to update and revise manuals so often fail.

Not only is there a continuing struggle to recognize and write up the needed changes; not only is there an eternally frustrating attempt to identify all the people and places that should receive the supplements and updates; there is also a vast set of random and perverse forces that conspire to misdirect and distort the effort. Mail systems, private or public, make errors – even if they are electronic or facsimile systems. Also, the recipients tend to misplace, misapply, misread, and otherwise abuse the bulletins. In how many manuals, for example, are all the supplements still wrapped in a paper strip, waiting to be incorporated?

The net effect is that every supplement to a manual – even though its purpose is to produce current, consistent documentation – may double the number of versions in circulation. (Some people receive the supplements; some don't.) When the original manual appears, there is only one version in circulation. (Not counting, of course, the unofficial versions extracted and created by industrious users.) With each added supplement, the number of alternatives doubles.

So, two supplements yield four versions. Four supplements yield 16.

After 10 supplements there could be 1K versions: 1,024!

This discussion is not intended to be humorous. Anyone who has tried to distribute corrections and updates to a large set of operators or customers knows that every possible misuse, misplacement, and mismanagement of the documents will, in fact, occur.

Incorrectly addressed materials disappear; correctly addressed materials are nevertheless mislaid. Materials that supersede older versions are stored in a desk, while the obsolete version remains in force at the terminal.

Although nothing can prevent completely this proliferation of misinformation, several measures can ameliorate it:

- Limiting the number of supplements and releases, keeping them in large batches, will reduce the noise in the documentation channel. Releasing these batches on a regular schedule solves another serious problem as well; it lets the users be confident that they have received *all* the supplements. When books are updated irregularly, the user is never sure.

- Putting as much documentation as possible into the system itself — thereby reducing the quantity of obsolete "hard copy" — will contain the problem.

- Limiting the updates to a single, authorized source will help.

- Requiring technical specialists to review and approve the documentation before it is sent will reduce the quantity of updates and corrections, especially the corrections of the corrections.

11.4   Can Old Manuals Be "Modularized"?

Few documents start from scratch; usually there is an old manual to be incorporated or updated. On occasion, an old book can be recast into a "structured format," but the format may be a little more than cosmetic, not really providing the benefits of a brand new, tested, modular manual.

Often, the assignment for newly hired documentors is to finish, update, upgrade, revise, or otherwise "clean up" some unacceptable publications. They are asked to begin long after the time when most of the document design decisions I have been talking about should have been made.

What about these existing documents? Can documentors charged with the task of editing or revising old manuals make use of this structured approach? Can an inaccessible, unreliable manual be made more usable?

Perhaps. The editors at Hughes Aircraft, when they first publicized their method of modular publication (the STOP technique) reported that they were able to recast old documents into the new two-page format. Partly, their success was due to the nature of the publications they were working with, many of which were already equal mixes of text and exhibits.

Given the right publication, the process can be almost fun. All the pages of the existing document are laid end-to-end, and the team of designers goes through the text and pictures marking off module-sized chunks of material, writing new headings or headlines, occasionally—

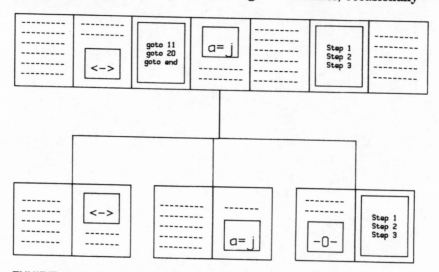

EXHIBIT 11.4   Recasting an Old Manual

but rarely—even rearranging the sections, or moving an exhibit from the appendix to the text.

Whether this a good idea depends on several factors. As already mentioned, some publications, which are closer to this format than others, call for less complicated reworking. Some manuals, because they are filled with good writing and pictures, are especially worth saving, and they justify the effort. But many documents (like many old programs) are obsolete and clumsy. Trying to save these is little more than yielding to the too common myth that it saves time and money to reuse existing material instead of creating new things.

Again, some old manuals lend themselves quite easily to "modularization." I have even seen people "back out" a storyboard from an existing manual that never had one in the first place. In many cases, though, **the effort to recast and retrofit an old publication is greater than the effort to generate a brand new one.**

But, even when the physical structure and format of the old book lends itself to recasting in the more readable modular format, please remember that modularity is not just an attractive way of presenting material. It is not just cosmetic.

Modular design, whatever its esthetic benefits, is also a way of assuring that the documents are maintainable and modifiable. A modular format imposed after the fact may improve appearance, but it might not significantly improve the maintainability and reliability of the book that has been revised.

I tend to agree with Yourdon and Constantine in their discussion of recasting old computer programs:

> It is all but impossible to simplify significantly the structure of an existing program or system through after-the-fact modularization. Once reduced to code, the structural complexity of a system is essentially fixed. It is thus clear that simple structures must be designed that way from the beginning.—*Structured Design* (Englewood Cliffs, NJ: Prentice-Hall), 1979, p. 35.

Recast and retrofit old manuals if you like. But do not miss the opportunity to engineer a publication from its inception. The differences in effectiveness and cost-effectiveness are remarkable.

The Future of
User Documentation

12

Documentation Without Books

12.1   Will There Be Paperless Systems?

The quantity of paper documentation seems to be decreasing—although one cannot know for sure. User information is, increasingly, stored and used without paper publications. More important, though, is the evolving nature of on-line systems, which, with each generation of improvements, seem to need less and less collateral user support.

A great and growing proportion of the people who use computer technology do not like books and do not use them well. At one extreme is a legion of clerks and data entry personnel, who, though able and resourceful, lack the skills they need to find things in a complicated manual. At the other extreme are the executives and managers who are too busy or impatient to read anything longer than three paragraphs and who are accustomed to getting their questions answered by other people.

Manuals also can be clumsy, especially the encyclopedic kind. Surprisingly few work stations have bookshelves; fewer have a place where a manual might be opened and read. And the task of working at a system with two manuals balanced on one's lap occurs far more often than is ergonomically safe.

Paper information products are also among the hardest to keep current. "Hard copy" (aptly named) does not let itself be changed nearly as readily as information stored on disks or tape. And old, obsolete publications can be so tenacious that they continue to be used after they have been superseded several times.

Even though there are several advantages to conventional manuals— especially for those of us who review and edit them—still, getting rid of paper documentation is a legitimate objective for some installations. And even if there is never a completely paperless system, there can surely be systems with very little of the stuff.

There are two broad strategies for achieving the objective. The more difficult strategy—ultimately the more useful—is to design systems in such a way that users need relatively little instruction after their initial orientation and, also, to assure that an operator or user will hardly ever have to look up anything. The key to both is to devise screens that communicate in plain English, especially in their prompts and messages.

A screen that prompts INSERT FMT DSK needs more documentation than a screen that prompts INSERT FORMAT DISK, which in turn needs more than INSERT THE FORMAT DISK FOR THE CHART YOU WISH TO CREATE. (Better still a screen that tells you which Format Disk goes with the Chart you selected; still better, a system that doesn't require you to choose, find, or insert any disks at all after you've logged on.) Furthermore, if the user has to look up a table of format disks, having the table on the screen further reduces the need for paper.

The most common reason to look things up is in response to coded or cryptic messages. ERROR 17 needs more support than SHOW/ WRITE ERROR, which still needs more than:

YOU MAY NOT **CHANGE** THE RECORDS IN A FILE
UNTIL AFTER YOU **SHOW** THE FIRST RECORD
IN THAT FILE

(And, of course, the most effective strategy is to revise the system so that this particular error either corrects itself, or cannot be made.)

Systems that tolerate errors usually need less paper. Systems that do not clutter their screens, that do not rely on abbreviations and compressed words are similarly effective. Indeed, one of the best ways to improve a system is to **revise any procedure that needs considerable explanation until it is self-explanatory.**

The other strategy for reducing paper, again, is to deliver the words and pictures electronically. Nowadays that almost always means through the video display of the system itself. In this case, the two most popular techniques are the interactive instructional program, usually called a "tutorial" in the software industry, and the HELP facility, a series of screens that present instructional or reference material to the user who asks for it.

Note that there is nothing new in the tutorial or the HELP screen. The tutorial is a form of programmed textbook, and the HELP facility is just a reference book with a computer program for an index. They are not new, but, in many cases they are a better way to do what must be done. In fact, when tutorials and HELP programs are written properly, they can have all the documentation virtues described in this book and completely eliminate the loops and branches that make manuals hard to use. In other words, well-designed on-line documentation can eliminate the need for most book skills.

The irreducible minimum for paper documentation is an Installation and Startup Guide. If all the user documentation is *in* the system or *on* the system, the only external manual we need tells us how to get the system going – at which point the prompts instruct us and the HELP screens solve our problems.

But just as there are tradeoffs between information products and information services, so are there tradeoffs between books and on-line information products. There is no *inherent* superiority in either. Without the right analysis, design, and testing, on-line documentation can be as hard to use as the most complicated manual.

12.2   Putting User Information On-Line

Although the material printed in user manuals can usually be put on-line as well, it should be clear that the mere use of the video display terminal cannot guarantee usability.

The most important thing to remember in putting user documentation on-line is that there is no magic in doing it that way. Until the advent of the personal computer, the most familiar example of words on screens was *microfilm*. And note that nearly every student, researcher, or parts department manager who has to use microfilm finds it awkward and unpleasant!

As is often the case with new technology, the principal advantages of on-line documentation are also its disadvantages. The compactness of the conventional video screen, with its 25 × 80 array of characters, forces the writer to be concise and to package information in very small modules. The fact that the user can see only one screen at a time ("windows" will be discussed in a moment) compels the documentor to integrate the material. This makes on-line instructions perfect for the neophyte reader, the person with limited book skills.

On the other hand, though, the compactness of the screen makes the on-line module too small for much explanation or demonstration. For, as hard as it is to read two pages at once, it is nearly impossible to read two on-line "pages" at once. Even the use of "windows"—texts that pop up in a morticed rectangle on the screen—does not help. What fits in a window is even less than what fits in a single screen. And putting three or four windows on a single screen, while it may make for an interesting demonstration at the computer show, usually generates more confusion than clarity.

These objections are, of course, bound to lose their force. There are already several "full screen" video displays with more than 50 rows of characters, and there is already window software that can show more than one of these full screens at once on a single tube. Eventually, this technology will be nearly commonplace.

Until then, the inescapable fact is that on-line screens are going to be much smaller than ordinary pages. It will still be possible, without clutter, to put three times as much on a two-page spread. And if it matters that a concept or process can be seen synoptically, with all the text and exhibits in one place, then on-line documentation will have limited applications.

For the most part, on-line documentation is to help beginners get going. That is why sophisticated products, like MicroPro's WordStar®, allow users to limit or shut off its on-screen reference. As Exhibit 12.2

```
   B:DOC12.2  PAGE 1 LINE 1 COL 01            INSERT ON  LINE SPACING 2
                < < <    M A I N   M E N U    > > >
   --Cursor Movement--   | -Delete- |  -Miscellaneous-   | -Other  Menus-
 ^S char left ^D char right| ^G  char | ^I Tab   ^B Reform | (from Main only)
 ^A word left ^F word right| DEL chr lf| ^V INSERT ON/OFF  | ^J Help  ^K Block
 ^E line  up  ^X line down | ^T word rt| ^L Find/Replce again| ^Q Quick ^P Print
      --Scrolling--       | ^Y  line | RETURN End paragraph| ^O Onscreen
 ^Z line down ^W line up  |          | ^N Insert a RETURN  |
 ^C screen up ^R screen down|         | ^U Stop a command  |
 L----!----!----!----!----!----!----!----!----!----!------------------R
 -------------------------------------------------------------------------
 12.  Documentation Without Books
 -------------------------------------------------------------------------
 12.2 PUTTING USER INFORMATION ON-LINE
 -------------------------------------------------------------------------
 Although the material printed in user manuals can usually be put on-line

 as  well,  it  should  be clear that the mere use of the  video  display

 terminal cannot eliminate strategic, structural, or tactical errors.  In

 some ways,  the screen has limitations that make it harder to use than a

 book...only some of which will be corrected by new communication methods
 1HELP    2INDENT 3SET LM 4SET RM 5UNDLIN 6BLDFCE 7BEGBLK 8ENDBLK 9BEGFIL 10ENDFIL
```

EXHIBIT 12.2   On-Line Information (Photocopy of Screen)

shows, if the user asks for the highest level of "help" in WordStar, the screen will fill with so much instructional material that there is hardly any room to see the document. That is why WordStar lets you remove all that help and also why most people who use the product simply refer to a reference card for the commands.

Better video screens and new software may prove all these complaints to be short-sighted. Larger and more readable displays will change the constraints. Also, dedicating a *second screen* to documentation would help, especially if the second screen were a high-resolution television device able to project fully defined movies and photographs.

No matter what the technology does, though, the user documentation issues remain the same. For, just as it is possible to type dumb sentences on smart typewriters, it is possible to write harmful information in Help messages. On-line documentation is merely a *medium*, a way of delivering to the users what they want and need. It must be planned and designed as carefully as — more carefully than — traditional manuals.

13

Afterword: Documentation in the Fifth Generation

13.   Afterword: Documentation in the Fifth Generation

The next generation of hardware (with its superior memory and speed) and software (with its superior "intelligence") will reduce the need for traditional user documentation. Eventually, the skills we now use to generate user documentation will be used for other things—mainly inventing applications and making the communication between people and computers more productive. At the same time, this new technology will also be used to *write and draw* exceptionally good user documentation.

The easier a thing is to use, the easier it is to write the user documents. In cases of marvelous design and engineering, the need for documentation approaches none. Only when the developer adds a new feature or function (color choices for your accounting software, equalizer settings for your stereo, memory for your telephone) is the technical writer called into action.

Does this mean that professional documentors and technical writers are going to be obsolete only a few years after the universities have begun to train them? Of course not. This new profession will not disappear so quickly.

For one thing, even if manual writers become obsolete in principle, we still can expect them to keep working for a while. In data processing, people keep doing obsolete things for years and years.

At the cutting edge of computer technology, though, conventional manuals will soon be less needed. Systems will be so "transparent" to their users that there will be little "human interface." And at those times when the system and user communicate with each other, the clarity of the computer's prompts and its ability to infer the intentions of the user will reduce the need to "support" the user.

"Support," which officially means helping the user get benefit from the product, is really more often an industry euphemism for teaching the user to adapt to the product. What will software writers do when the product learns to adapt to the user?

It is obvious. They will design the human and intellectual components of the systems, as well as writing the on-screen prompts and on-line reference tools. **In the Fifth Generation and beyond, the skills of a professional communicator will be more critical to the success of systems than the skills of an electronic or software engineer.** Furthermore, those skillful writers who also understand the technology and its potential may become the leaders of development, instead of the ones who just try to clean up the mess after the technical troops have marched through.

Also interesting to contemplate is the effect of Fifth Generation technology on the writers and documentors themselves. For more than five years now, I have been advocating to my clients that everyone in any organization who writes should do so directly into a word processor – not just to save money on typists, but to improve the performance of every writer. What becomes clear to nearly everyone who has made the switch from pens and typewriters – especially to professional writers – is that word processing raises the quality of one's work. It makes every editing and revising task easier, thereby encouraging editing and revising; it narrows the gap between the cerebral word and the physical page; it lets us cut and paste with abandon. Furthermore, when used in connection with telecommunications, the new technology allows teams of documentors to "network" on projects, sharing current knowledge and enhancing each others' writing and presentation skills.

What is emerging is a new era of "author power," in which the originator or manager of a document will be able to control content, page layout, graphics, typesetting, and distribution. In which text and pictures will be shuffled in and out of data bases and across the nodes of communication networks. In which technical changes will be communicated nearly automatically into procedures documents, and from there to waiting tutorial programs.

Nearly all these improvements in the technology of writing, drawing, and publishing tend to reduce the cost and effort of changing a manuscript. Revisions that, only a few years ago, would have been prohibitively expensive and time consuming are now cheap and fast. This means that, in a few years, some of the more defensive formal methods advocated in this book will be appropriate only for relatively large documents. For small documents, it will cost no more to build them, test them, and throw them away than it would to build and test models.

Gradually, the new technology will not only enhance the quality and usability of our documentation, but also reduce its short-term cost and production time. In effect, this will overwhelm two of the main enemies of quality documentation: myopic concern with immediate expense and impatient preoccupation with deadlines. In other words, even though usable user documentation always pays for itself, in the Fifth Generation the payback period will be dramatically shorter.

In sum, the next generation of knowledge processing technology will give us the ability to write and draw what we will. Our legs will keep pace with our desires. Let us hope that we shall use this extraordinary power to pursue what William Zinsser calls the three cardinal goals of good writing: clarity, simplicity, and humanity.

Appendixes

A.   Illustrative Tables of Contents for "Usable" Manuals

The exhibits demonstrate the principles of outlining advocated in this book. (Outlines, of course, become Tables of Contents.)

EXHIBIT A.1   Operator Guide for a Line Editor Called LINE-ED

---

1. How to Read a LINE-ED Manual
2. The Two Functions: Entering and Editing
3. How to Enter LINE-ED: The LEDIT Instruction
4. How to Write a LINE-ED Command
   4.1 Controlling the Cursor/Pointer
   4.2 Entering New Data (Input)
   4.3 Inserting Additional Data
   4.4 Merging Two Sets of Data
   4.5 Locating Past Data
   4.6 Deleting Unwanted Data
   4.7 Replacing Data
   4.8 Making "Global" Changes in Files
5. How to Manage LINE-ED Output
   5.1 Formatting Input or Output (Three Methods)
   5.2 Specifying Print Protocols
6. How to Recover Deleted Items
Appendix: LINE-ED Messages

EXHIBIT A.2   Office Supervisor's Guide to Word Processing System

---

1. Lines of Authorization in the WP Center
   1.1 Authority: Who May Approve a Job
   1.2 How to Assign Job Priorities
   1.3 Table of Organization
   1.4 Tables of Duties and Responsibilities
   1.5 Eight Preconditions for the Use of WP Facilities
2. How to Start the WP Equipment
   2.1 Starting the Main Processing Units
   2.2 Starting the Printers
   2.3 Opening the Communications Channels
3. Level I Jobs: Basic Correspondence
   3.1 Defining a Document File
   3.2 Entering a Text
   3.3 Printing a Review Copy
   3.4 Editing the Text
      3.4.1 The Twenty Most Common First-Draft Errors
      3.4.2 The Three Most Difficult Revisions
   3.5 Printing the Finished Copy
   3.6 Sending the Finished Copy Through the Electronic Mail
4. Level II Jobs: Complicated and Technical Documents
   4.1 Assembling Documents from Older Documents
   4.2 Merging Document Variables
   4.3 Performing Arithmetic with the WP Software
   4.4 Generating a Mailing/Distribution File
   4.5 Interpreting Ambiguous Input (Default Rules)
5. Policy: Logging and Storing of *All* Documents
6. Policy: Protecting the Confidentiality of Our Clients
7. Policy: Resisting Pressure from Originators and Managers

EXHIBIT A.3 Manual for Creating a Special Purpose Phone Network (Excerpt)

Chapter 3. Building and Verifying a TERRITORY-MAP
- 3.1 Each TERRITORY-SET is a Telephone Data Base
- 3.2 How to Start a New TERRITORY-SET
- 3.3 How to Enter the Data for One Sector's TERRITORY-SET
- 3.4 How to Transfer Set Data from the AZ-60 to the AZ-190
- 3.5 How to Create an Auxiliary Data Base Called INTERIM
- 3.6 How to LOAD Data from INTERIM to TERRITORY-SET
- 3.7 How to Get Access to GRAF-MAP
- 3.8 Five Requirements for All Maps
- 3.9 How to "Introduce" a Map to the System
- 3.10 The Most Efficient Procedure for Entering Map Data
- 3.11 How to Use the ZOOM Feature of GRAF-MAP
- 3.12 How to Manipulate a Map
  - .1 Adding Buildings and Nodes
  - .2 Changing Buildings and Nodes
  - .3 Adding Terminal Data
  - .4 Changing a Terminal Profile
  - .5 Modifying the Terminal/Site Matrix
  - .6 Adding a Segment Between Two Nodes
- 3.13 How to Validate Entered Data
- 3.14 Review: The TERRITORY-SET Checklist

EXHIBIT A.4 User Guide for the E-POST Electronic Mailing System

1. Conventions Used in This Manual
2. Three Reasons to Use E-POST Instead of Ordinary Mail
3. Knowing When You Have Received an E-POST Message
4. Getting Access to E-POST
5. Getting Help from E-POST
6. Using the E-POST
7. Using the E-POST Index
   - 7.1 Getting Access to the Index
   - 7.2 Searching for U-Names by Last Name
   - 7.3 Searching for U-Names by Account Number or ID
   - 7.4 Locating U-Names on Other OGR Computers
8. Using Distribution Lists: Overview
   - 8.1 Setting Up a Distribution List
   - 8.2 Changing a Distribution List
   - 8.3 Directing E-POST Messages through a Distribution List
9. Printing E-POST Messages: Three Methods
   - 9.1 Printing the Screen
   - 9.2 Printing with the STORE Facility
     - 9.2.1 Saving an E-POST Message as a STORE
     - 9.2.2 Printing the STORE File
   - 9.3 Printing with the Word Processing Facility
     - 9.3.1 Saving the E-POST Message as a Document
     - 9.3.2 Sending the Document to SCRIBE-15
     - 9.3.3 Printing the Message by its SCRIBE-15 Document ID
10. How to Send an Existing File as an E-POST Message
11. How to File E-POST Messages in Secondary Indexes
12. E-POST Commands

B. Glossary of Selected Terms Used in This Book

The table below defines some of the important terms used in this book.

Term	Definition
accessibility	the ease with which information can be found or extracted
artistic stereotype	a method of writing in which most of the effort is in the draft and relatively little in analysis and design
audience	a group of readers with common interests (similar tasks) and common background
availability	the presence or absence of a document
demonstration module	part of a manual that teaches an entire process to an experienced reader
documentation set	a plan defining all the user manuals and other information products associated with a system
engineer stereotype	a method of writing in which most of the effort is in analysis and design and relatively little in the first draft
GOTO	an unconditional branch; in manuals, entering or exiting a module in the middle
headline	a thematic or substantive heading, associated with one module of documentation
instructional module	part of a manual that teaches one new task or idea to a neophyte reader
maintainability	the ease with which systems or manuals can be debugged, repaired, and modified
model	representation of a system or product, used to facilitate testing
module	small, functional, independent entity in a system or document
motivational module	a part of a manual that gets readers to perform a task they are reluctant to perform
readability	the ease with which a passage can be read, often expressed in grade level of difficulty
redundancy	deliberate repetition and duplication, meant to reduce the burden for the reader and offset the effects of noise and distraction
reference module	a part of manual that serves as auxiliary memory for the user to "look up"
reliability	absence of interruptions and failures
storyboard	a working display showing the specs for each module in a manual, a model for the emerging publication
strategic error	failure to develop the right mix of publications and information products

structural error	failure to organize the contents of a publication into the most usable sequence
structured	of a process, developed through top-down analysis and modeling; of a product, organized into modules and the links that couple them
structured outline	a list of headlines naming each module in a planned manual
suitability	the degree to which a manual fits the interests and supports the tasks of the user
tactical error	failure to edit drafts for clarity and readability
task-oriented	of documentation, defined so as to support users in precisely what they do
topic/audience matrix	an array of topics to be documented and users/readers to be supported; used in defining the mix of documents
usability	the ease with which a system, product, or manual can be used
usability index	the more often the intended reader must skip, branch, or detour, the less usable the book
user documentation	all the information products devised to help users adapt to their technology

C.   Books and Resources for Documentors

Even though the study of user documentation is a relatively new field, it is not brand new. There is a small but powerful library of books and periodicals that will be immensely helpful to a neophyte documentor—or even to a veteran.

Before presenting some recommended books in the usual alphabetical sequence, it is useful to point out an especially valuable one:

> Sandra Pakin & Associates. *Documentation Development Methodology.* Prentice-Hall, 1984.

This unusually practical work is just the thing to get a novice writer started. And, indeed, companies without publication standards can even adopt the contents of this book as a Standards and Procedures manual for the organization.

In addition to the Pakin work, here are some other useful books about user documentation:

> d'Agenais, J., and J. Carruthers. *Creating Effective Manuals.* South-Western Publishing Company, 1984.
>
> Grimm, Susan. *How to Write Computer Manuals for Users.* Lifetime Learning, 1982.
>
> Horn, Robert. *How to Write Information Mapping.* Information Resources Inc, 1976.
>
> Kelly, Derek. *Documenting Computer Application Systems.* Petrocelli Books, 1983.
>
> Matthies, Leslie. *The New Playscript Procedure.* Office Publications, 1977.
>
> Rubin, M. L. (ed.). *Documentation Standards and Procedures for On-Line Systems.* Van Nostrand Reinhold, 1979.
>
> Schoff, G., and P. Robinson. *Writing & Designing Operator Manuals.* Lifetime Learning, 1984.
>
> Zaneski, Richard. *Software Manual Production Simplified* Petrocelli, 1982

Many of the people assigned to work on user documentation are new to the ranks of technical and professional writing. The titles below are among the best general works on how to write clearly, especially about technical and scientific subjects.

> Brogan, John. *Clear Technical Writing.* McGraw-Hill, 1973.
>
> Michaelson, Herbert B. *How to Write and Publish Engineering Papers and Reports.* ISI Press, 1982.
>
> O'Rourke, John. *Writing for the Reader.* Digital Equipment Corp. No. DEC-OO-XWRDA-A-D, 1976.
>
> Skees, William. *Writing Handbook for Computer Professionals.* Lifetime Learning, 1982.

Strunk, W., and E. B. White. *The Elements of Style*, 3rd ed. Macmillan, 1979.

Van Duyn, Julia. *The DP Professional's Guide to Writing Effective Technical Communications*. Wiley, 1982.

Weiss, Edmond. *The Writing System for Engineers and Scientists*. Prentice-Hall, 1982.

Often, the people writing user publications are also involved in writing a full range of other system-related documents, mainly plans, specifications, and reports. The books below discuss the documents needed in the management and administration of computer projects.

Harper, William. *Data Processing Documentation*, 2nd ed. Prentice-Hall, 1980.

London, K. *Documentation Standards*, 2nd ed. Mason & Lipscomb, 1974.

Long, Larry. *DP Documentation and Procedures Manual*. Reston, 1976.

National Bureau of Standards/Institute for Computer Sciences & Technology
Computer Model Documentation Guide (500-73)
Proceed's of the NBS Software Documentation Workshop (500-94)
Guidelines for Documentation of Computer Programs and Automated Data Systems (FIPS PUB 38)

Poschmann, A. *Standards and Procedures for Systems Documentation*. AMACOM, 1983.

Sides, Charles H. *How to Write Papers and Reports About Computer Technology*. ISI Press, 1984.

In addition, I recommend that people without a DP or computer sciences background develop the habit of reading the "In-Depth" section of *Computerworld* each week; some of the most creative thinkers in the industry present their newest ideas and preview their partly written books in these pages.

Articles about user documentation pop up everywhere, even in the Sunday newspapers. But certain periodicals deal regularly with the subject:

FOLIO, published by Pakin & Associates in Chicago

(The SIGDOC Newsletter), published by the Special Interest Group on Documentation of the Association for Computing Machinery (ACM) in New York

Technical Communication, the Journal of the Society for Technical Communication, Washington D.C.

IEEE Transactions on Professional Communication

"Simply Stated," published by Document Design Center, Washington, D.C.

Probably the most readable collections of papers and articles are the annual *Proceedings of the International Technical Communications Conference*. These anthologies, available through the Society for Technical Communication, are the most provocative and useful books on technical communication, in general, and user documentation, in particular.

D.   Software Tools for Documentors

Writers of user manuals should immediately acquire the equipment and skills they need to work directly at their own word and information processing stations. This will not only improve the quality of their documentation, but also give them revealing insights into the problems of users, systems, and publications.

Word and Text Products

A *full-featured word-processing package* that allows the complete range of text movement (cut-and-paste) and formatting options (page layout). Some of the more advanced features in WP programs are footnoting, nonstandard page-numbering, and print spooling (the ability to print one document while editing another one).

A *typesetting interface* that converts the files from your word processor into magnetic instructions for photocomposition, without the need to have the printer re-enter the text.

A *lettering program* that allows you to print or plot headings and exhibits with oversized characters in assorted fonts. (These programs are also available as chips that can be inserted in printers.)

An *outlining tool*, one of a growing number of programs that lets the author play with ideas and organize them into alternative hierarchies and sequences.

A *text-management facility* that lets you code your text and organize it into a data base, so that documents can be "written" merely by retrieving a sequence of identified passages or "modules."

A so-called *authoring tool or language*, a software product to help writers create computer-based training programs or on-line tutorials.

Editorial aids, programs that check spelling and even, in some cases, flag style and usage problems or calculate readability scores.

Graphics and Art Products

A *business chart package* that can be used to draw simple line, bar, and pie charts from statistical data.

A *sign-making package* designed to make title pages and other presentation materials (including slides and transparencies) containing mostly letters.

A *diagram tool* designed for flowcharts, data flow diagrams, organization charts, GANTT charts, and similar diagrams that are used frequently in business and data processing.

Drawing and painting programs that allow the creation of freehand drawings that are stored magnetically and rendered through the printer or plotter. Often, these products contain a repertoire of standard symbols that can be assembled into various configurations.

A *"clip art" library* of symbols and cartoon illustrations that can be printed or plotted in business communications, usually as an extension of a drawing or painting program. Often, a disk with hundreds of pictures costs less to acquire than having one or two drawings rendered by a commercial artist.

Graphics enhancers, programs that enrich and improve simple business charts and diagrams, often by adding a simulated third dimension.

Utilities and Productivity Products

A *project management package* that allows the manager of publications to schedule and budget the development of the manual or other information product. If manuals are designed and produced in the modular way, then all the features of a project management system can be exploited: critical path scheduling; resource leveling; long-term estimating . . .

A *key defining program* that allows the documentor to change the function of the keys on the keyboard. Programs are available that allow you to redefine any key (or even change your whole keyboard to a nonstandard one), or, more important, make a short series of keys represent as many as 50 or 60. Whether for creating "macros" (long procedures that run on one, two, or three keystrokes) or putting recurring phrases into the lexicon of the computer (names and terms that are used repeatedly), these utility programs can save countless hours of tedious effort.

Finally, a *spreadsheet and file manager*. These most popular business programs can be extremely useful to a documentor as well. Spreadsheets can be used not only for scheduling and financial planning, but even to generate the Topic-Audience Matrix used to analyze documentation needs. File managers (or data base management systems) are essential in keeping track of documents, especially of their maintenance needs. Further, most file managers can be programmed for the storyboard process, as tools for storing, printing, and manipulating module specs.

Index